# An Introduction to 3D Printing

Victoria Zukas, MPS
Jonas A. Zukas, PhD

D1059835

An Introduction to 3D Printing
Copyright ©2015 Victoria Zukas and Jonas A. Zukas

ISBN 978-1622-878-96-3 PRINT
ISBN 978-1622-878-97-0 EBOOK

LCCN 2015938362

May 2015

Published and Distributed by
First Edition Design Publishing, Inc.
P.O. Box 20217, Sarasota, FL 34276-3217
www.firsteditiondesignpublishing.com

Cover Design: Victoria Zukas

# About the Authors

Victoria E.L Zukas received her bachelor's degree in Interactive Media and Game Development in 2009 from Worcester Polytechnic Institute. She completed her Masters of Professional Studies in Digital Media in 2013 at Northeastern University. She has spent her time since then working as a Freelance Graphic Designer on a number of projects. Her background is mainly focused on creating 3D models for video games. Her published game BLASTiators can be found on the Google Play store.

Jonas A. Zukas, received his doctorate in engineering mechanics from the University of Arizona. He is widely known for his work in numerical modeling of the behavior of structures and materials at high rates of strain. He has co-authored and edited a number of books and conference proceedings and is the author of the monograph Introduction to Hydrocodes published by Elsevier. He is co-developer of the ZeuS code for the analysis of impact phenomena and was a Fellow of the American Society of Mechanical Engineers, a Senior Member of the American Institute of Aeronautics and Astronautics and a member of the American Academy of Mechanics.

# Acknowledgements

First and foremost we thank those individuals and organizations which permitted us to use their illustrations in the book. These have helped to illustrate points that we had made in words and thus added to the clarity of the discussions. Their individual contributions are noted throughout the text.

We thank Muriel Russell, Victor Russell and Joshua Ouimette for their diligent review of parts of the manuscript and their valuable suggestions which greatly improved readability.

We also thank Mark Sivak and Tim Carroll for sharing with us their advice and knowledge on 3D printing.

# Introduction

When undertaking the production of a book on three-dimensional (3D) printing, it behooves the authors to declare the causes that impel them to do this.

The technology which enables printing in three dimensions, under one name or another, has been around since the 1980's. Had this book been written at that time, its focus would have been on the novelty of 3D printing and, perhaps, a cloudy vision of its future. 3D printing was the topic of graduate theses and dissertations. Specialists in universities and a handful of commercial facilities experimented with 3D printers for rapid prototyping. Existing computer-aided-design (CAD) software was being adapted by specialists primarily to create models for 3D printers. A lot of the work was seat-of-the-pants engineering, developing the necessary hardware and making software adaptations as situations required them. In due course, fascination with the concept grew, 3D printer kits became available and a small army of hobbyists and do-it-yourselfers joined the field.

So what has changed? Some 250 – 300 3D printers are currently available on the market. They range from industrial machines the size of moving trucks that can produce aircraft components and automotive parts with dimensions measured in tens of feet, to desktop units for the home with build platforms measured in inches. Software packages are now readily available aimed at both engineers and artists. 3D printing has made dramatic, even life-saving, contributions to medicine. A 3D printer is now on board the International Space Station. Fashion designers use 3D printing to create jewelry of all descriptions, footwear and designer dresses. In the 1980's, if you printed anything it was made of plastic, the only material available at the time. Today the range of materials that can be used includes plastics, metals, concrete, ceramics and even food.

Today, because of technical improvements and lower costs, 3D technology is available to everyone, not just specialists. Accordingly, information about 3D print technology has also migrated from literature aimed at specialists to articles in popular media aimed at, and accessible to, a curious reader whose first reaction to the mention of 3D printing might have been "What?" That question is now answered in mass media – newspapers, magazines, television shows - as well as social media – YouTube, Twitter, an assortment of blogs and newsletters devoted to 3D printing. Any curious individual can now learn about the technology and use it to satisfy their own needs.

Because so much information is now available, the challenge is to find the subset that is useful to an individual or organization from the vast pool that has been generated. Without a little guidance, one can easily spend months looking for and making sense out what is available in order to answer the questions one needs answered.

This book is aimed at an audience consisting of two kinds of readers. The first is people who are curious about 3D printing and want more information without necessarily getting deeply into it. For this audience, the first two chapters will be of greatest interest. They provide an overview of 3D print technology. They also serve to take the confusion out of the jargon and make sense out of such shortcuts as SLA, FFM, FFF, FDM, DLP, LOM, SLM, DMLS, SLS, EBM, EBAM, CAD and others. They describe the basic processes, the materials used and the application of the technology in industry, space, medicine, housing, clothing and consumer-oriented products such as jewelry, video game figures, footwear, tools and what must now seem like an infinity of bunnies, eagles and busts of Star Wars and Star Trek figurines in a dazzling array of colors.

This book also addresses the needs of people new to the field who require information in a hurry. Chapter 3 serves as a guide to generating a 3D model by reviewing scanning methodology, the various types of software available to create a model and the steps needed to insure a useful printed object from the 3D model. The chapter has numerous references which, together with the information in the text, will help one find quickly any additional information available on the internet.

There is a steep learning curve associated with the software used to generate 3D models. Chapter 4 addresses the needs of people who are curious to try the technology but, if they use it only sparingly, may not want to make the investment in either the hardware or the time it would take to learn the software. We review some of the printing services available and the model repositories which provide models that can be downloaded for a fee and sent on to a printing service. A discussion of the issues involved in deciding whether to buy a 3D printer or use the available services to do 3D printing without a 3D printer is included as well.

If you are brave enough to try to design your own object, Chapter 5 is an exercise which walks you through the characteristics of an available software package and the steps required to design a practical object, in this case a screwdriver. Doing this exercise – not just reading through it – will give you considerable insight into the capabilities of 3D software packages and, we hope, build your confidence and encourage you to try again with an object of your own choosing.

There is one aspect of 3D printing that we do not address. You will find no mention of building printers from kits or do-it-yourself (DIY) projects. It is our opinion that the 3D printing industry is, at this stage, sufficiently advanced beyond the hobby stage that this aspect is best left to those who enjoy the hands-on experience. If this is something that interests you, we suggest that a good starting point would be the various publications and activities of Maker Shed, a division of Maker Media, Inc. Check out MAKE magazine and subscribe to the newsletter to stay informed on the wide variety of activities they sponsor.

# An Introduction to 3D Printing

Victoria Zukas, MPS
Jonas A. Zukas, PhD

# Table of Contents

# Chapter 1

## An Overview of Three-Dimensional (3D) Printing

### 1.0  Introduction

In the two-dimensional world of computing, you plug a printer into your computer, specify what it is you want printed, and press "Print".  When it comes to 3D printing, the situation is just a tad more complex.

The current state of 3D printing resembles the early days of aviation. The Wright brothers have flown their wood frame and canvas plane, powered by what today would be a lawnmower engine, a distance of 50 feet. The proof of principle has been established. Poorly functioning, inefficient designs have been weeded out. Now the race is on to build better, if not bigger, aircraft, but these are still variations on the Wright Brothers ideas. Services such as an early version of airmail and single passenger travel exist, but multi-passenger planes, the Pan Am Clipper ships, jet engines, supersonic flight and space shuttles are still far in the future.

We are at the beginning of yet another technological revolution. Where it will end is anyone's guess but given the wide variety of applications thus far, it is clear that this is not a fad or a tinkerer's toy.

Let's go back to 2D printing for a moment.  Printers have been part and parcel of computing for more than 70 years. Software has been developed and incorporated in computing systems that make it a straightforward process to obtain two-dimensional printed and graphic views of computational results. Graphs of scientific, business and financial analyses are readily generated and displayed.  Architectural drawings and  artistic designs are readily available and transferrable for business or pleasure. The hardware and software which accomplishes this have evolved, been integrated throughout computing systems and made accessible to specialists and the general public alike through user-friendly interfaces.

Add another dimension to this process and new vistas open up but at the cost of additional complexity. 3D printing does not yet have the benefit of seamless integration with its components. The "plug-and-play" era is still some years in the future. Hausman and Horne [1] compare the current level of sophistication in 3D printers as being close to the first automated looms in factories in the 1700's. A comparison can also be made with the early days of the automobile. An owner of a motor car in the late 19th and early 20th centuries needed not only know how to drive the vehicle but also be able to do a certain amount of tinkering with its innards, such as lubrication, oil changes, replacement of worn parts (spark plugs, filters, hoses). The owner had to be able to do minor repairs as garages in that era were few and far between. Current users of 3D print technology need to know a fair amount about the computer software used to generate the digital models as well as the eccentricities of the hardware that is used to create them. At the moment, the learning curve for this technology is steep. Judging by what has been accomplished thus far and its potential, it is time well spent.

## 1.1 The Basics of 3D Printing

The language of 3D printing is illustrated in Figure 1.1. The axes, x, y and z, which are depicted is the space in which we will work. The x-y plane is the base where we begin, the z-axis denotes height. Shrink the z-axis to zero and you have the paper that is generated by a 2-D printer. In the three-dimensional world, imagine a thin layer of material being deposited on the x-y plane. This layer has a small, but finite, thickness. Once this first layer has been deposited, add a small amount of material in the z-direction, building another layer upon the first one. Keep this up and soon you have a 3-dimensional structure. Figures 1.2 to 1.5 show some steps in the process. The object you want to "build" is created layer by layer to achieve the desired result. The structure, or object, you create is limited only by your imagination and your talent in manipulating the computer-aided-design (CAD) software. If this sounds deceptively simple, that is because we're deferring the complexities to further sections dealing with hardware, software, choice of materials and polishing the final design. Right now, just focus on the basic principles.

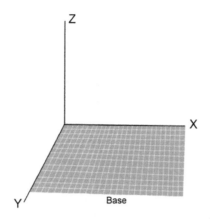

*Above: Figure 1.1 -- Three-Dimensional Coordinate System*
*Below: Figure 1.2 – 1.5 – 2D representation of the layering process in 3D printing*

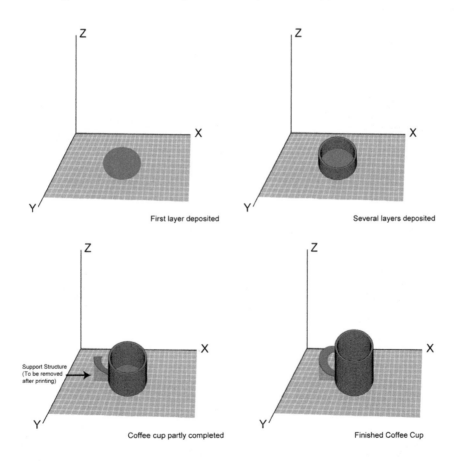

The process just described was originally referred to as additive manufacturing. Conventional manufacturing techniques start with a solid three-dimensional shape and cut away unneeded material until the desired object appears (subtractive manufacturing). Sculpting an object is a good illustration of this approach. The great sculptor Michelangelo started with large blocks of stone from which emerged masterpieces such as the Pieta and the statue of David. He was once asked how he created such masterpieces. Michelangelo is said to have replied that he sees the angel inside the solid block and removes stone until the figure emerges.

You can already see one advantage of the additive over the subtractive process. The additive process uses only the amount of material needed to achieve a desired result. Subtractive methods discard material while creating an object. The greater the cost of the material, the greater the advantage in minimizing its use.

The earliest use of additive manufacturing was in rapid prototyping [2]. Prototypes serve as preliminary models of stuff (cars, planes, machines, parts thereof) which allows the design to be viewed, tested, corrected, and later serve as a model for the final product that is to come. Prototyping has been a part of engineering design for a long time. Once computer-aided-design (CAD) software became available, designers used it to create models. This digitized version of a model was then sent to a machine which actually constructed the object.

Before the 1990's the different processes of additive manufacturing went under different names. By 1984, Chuck Hall had developed a process called Stereolithography (SLA). The basic process works like this [3]:

- A CAD program is used to create a model of an object
- A piece of software (digital slicing)chops the CAD model into thin layers, typically 5 – 10 layers/millimeter
- The three-dimensional printer contains a liquid photopolymer (resin). An ultra-violet laser "paints" or "cures" one cross-section at a time. On each layer, the laser beam traces a pattern.
- The printer platform drops down into the tank one layer thickness (a fraction of a millimeter). More resin is introduced and, on this new liquid surface, the laser retraces the pattern which now adheres to the previous layer.
- The process repeats until the entire object is cured.
- At the end, the excess print material and any support structures are removed. The finished object is cleaned of excess resin by immersion in a chemical bath and then cured in an ultra-violet oven.

Stereolithography can print almost anything that can be designed in a CAD program. The SLA process can produce objects with high detail and good surface finish. On the down side, support structures are required to prevent certain geometries from deflecting (or collapsing) due to gravity and to hold the 2D cross-section in place to resist lateral pressure from the re-coater blade. The supports are removed manually after printing. In addition, the machines need to be vented because of the fumes from uncured materials can be toxic. Also, since both the printers and the polymer resin are expensive, it is uncommon to see stereolithography machines anywhere but in large companies.

Chuck Hall also developed digital slicing and the STL file format [4], all still in use today.

By 1990, Stratasys, a company founded by Scott Crump, commercialized a plastic extrusion process called "fuzed deposition technology". In 1995, the Massachussetts Institute of Technology (MIT) developed a procedure which it trademarked with the name 3-D Printing. At this point, all the various additive manufacturing technologies retain their original names but are also collectively referred to as 3D printing.

3-D printing is more a system than a device, the printer being but one component.

- First, a digital model of the object to be created must be made with the help of CAD software or by scans.
- Next, the digitized model is translated (via more software) into a form readable by a 3-D printer (known as STL format). Additional software then "slices" the model into a number of layers depending on the accuracy required for the end product. The slicer output is a G-code which is then sent to the printer.
- The 3-D printer then creates a solid model, the process taking minutes to days depending on the complexity of the design, the accuracy (resolution) required, the size, the materials used, etc.
- The 3-D model is then reviewed by the designer. Often, changes need to be made to remove flaws, improve the look, remove support structures, enhance artistic aspects, etc. These changes are made by (you guessed it) additional software and the refined model again sent to the printer.
- This process is repeated until the developer is satisfied with the end result.

Figure 1.6 summarizes the process.

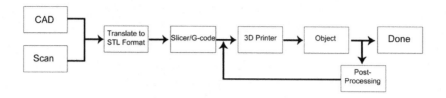

*Figure 1.6 -- The 3D Printing Process*

Not all 3D printers work the same way. Just as 2D printers come in inkjet and laser models, 3D printers operate under a variety of processes to achieve a common end result. There will be more on this later, but this is a good time to introduce you to the methodology and the industry jargon.

## 1.2 3D Print Methods and Materials

The main differences on the way layers are made to create an object from liquid materials are the following:

- Curing (hardening) a liquid material – this is the stereolithography (SLA) process already mentioned.
- DLP Projection uses a projector to solidify object layers one cross-section at a time rather than using a laser to trace the slices.
- Material jetting or polyjet matrix method uses a light source to solidify a liquid polymer. Object layers are formed by the emission of a liquid photopolymer from an inkjet-style multi-nozzle print head. After each layer is printed an ultraviolet light is used to solidify it before the next layer is printed.

In some printers, paper can be used as the build material, resulting in a lower printing cost. This is referred to as Laminated Object Manufacturing (LOM). In the 1990's, printers were marketed that cut cross-sections of paper coated with an adhesive using a carbon dioxide laser and then laminates the pieces together [5]. In 2005, Mcor Technologies, Ltd. developed a different process using ordinary sheets of office paper, a tungsten carbide blade to cut the shape and selective deposition of adhesive and pressure to bond the object [6,7]. There are several advantages to the use of paper printers:

- The printer uses standard office paper, readily available and cheaper than alternative printing materials, and ink
- Printing in full color is possible

- No chemicals are needed to dissolve support material and there are no toxic fumes to vent

Several examples of paper-printed objects are shown in Figures 1.7 – 1.8.

*Left: Figure 1.7 -- Architectural Model (Courtesy Mcor Technologies)*
*Right: Figure 1.8 -- A paper orange (Courtesy Mcor Technologies)*

Printers are also available that print laminated objects using thin plastic and metal sheets as the build material.

Material extrusion of Fuzed Deposition Modeling (FDM), also known as PJP, FFM and FFF, was developed in the late 1980's by Scott Crump and his company, Stratasys. With the FDM process, the object is produced by extruding small beads of material which harden immediately to form layers. A thermoplastic filament or metal wire that is wound on a coil is unreeled to supply material to an extrusion model head (Figure 1.9). The nozzle head heats the material and turns the flow on an off. An overview of 3D printing filaments in use is given by Ragan and Stultz [8]. The book by France [9] also contains a list of vendors that sell 3D printer filaments. Multiple colors are available (Figure 1.10).

*Left: Figure 1.9 -- FDM filament coil*
*Right: Figure 1.10 -- Multicolor FDM coils*

The FDM process is good for producing strong, complex, albeit low quality end-user products or prototypes. Unlike the SLA process, no post curing is required but support structures are required. FDM printers are environmentally friendly and thus suited for home or office use. Low end, economical machines are readily available. FDM is slower than other 3D print processes, produces rough surfaces (which can be polished or treated with a finishing coating), and does not have the high z-axis resolution of the SLA method [10]. Because of its accessible cost, the FDM approach is popular among hobbyists and home users. It is somewhat restricted in the shapes that may be fabricated. If the object has asymmetries or large angular deviations from the normal, a support structure must be included in the design. This can be removed after printing. Examples of FDM printers for industrial and home use are shown in Figures 1.11 and 1.12.

*Left: Figure 1.11 -- Stratasys Connex3 Multicolor Printer (Courtesy Stratasys)*
*Right: Figure 1.12 Cubepro Lifestyle Desktop (Courtesy 3DSystems)*

FDM is patented and trademarked by Stratasys. Other manufacturers therefore refer to the same process as "thermoplastic extrusion", "plastic jet printing" (PJP), the "fuzed filament method (FFM) and "fuzed filament fabrication" (FFF).

The utility of 3D printers is greatly expanded if they can be made to work with harder materials with high melting temperatures. To achieve this, the heated nozzle used to melt plastics is replaced by a much more energetic source such as a laser or an electron beam. The print processes for this purpose are referred to as:

- Selective Laser Melting (SLM)
- Direct Laser Metal Sintering (DMLS)
- Selective Laser Sintering (SLS)
- Electron Beam Melting (EBM) and Electron Beam Additive Manufacturing (EBAM)

Both prototypes and finished products can be 3D printed from powdered thermoplastics and metals that are cured (sintered) by a laser or an electron beam. Sintering is the process of making a powdered material into a solid by heating and/or the application of pressure without causing it to liquefy. In the Selective Laser Sintering (SLS) process, the material from which the object is to be made is introduced into the working area of the printer in powder form and cured. Powder is then added for the next layer and the process continues until the job is finished. At the end, the excess powder is removed and can be used for the next job. There are minor variations on this idea which are referred to as DMLS and EBM. The processes are variously referred to as SLM, EBM and DMLS (see above), some better suited for one material or another [11, 12].

Compared to other methods, SLS uses a wide range of material powders to create prototypes for new products as well as end products for customer use. The method also allows for the creation of complex shapes. The powder acts as a support during printing of the object which permits creation of shapes not possible with traditional manufacturing methods. The process is fast and economical.

At this point we should note that the terms "fast" and "economical" in the 3D printing industry are relative. Speed is dependent not only on the machine but also the size of the object being printed, the complexity of the object's design and the material(s) used in the process. When a cost comparison is made, it is usually between the cost of materials involved in 3D printing vs conventional manufacturing. The cost of the 3D printer may not be figured into the calculation. Certainly the designer's time in preparing the CAD model is not. Other considerations such as the time involved in printing an object (hours for something small, days to weeks for large complex objects) need to be considered as well. Some examples are given in the next chapter. On large jobs, companies simply let the machine run overnight so that it does not take work hours to print the object. In general, the time a 3D printer takes to print out various parts seems to be much less that the time required to create the parts with traditional manufacturing methods, at least in the prototype stage. Caveat Emptor (Let the Buyer Beware) is a good guideline when comparing costs.

On the down side, the finished object created via SLS tends to be rough and porous. This can be overcome with post-production coating and polishing. As compared to the SLA process, the detail of an object will not be as sharp and crisp. Finally, the advantage of being able to use a wide variety of materials, laser sintering printers, at this stage of their development, tend to be large, cumbersome, expensive and used mainly for industrial applications. That may be changing as recently Sintratec has introduced "the world's first desktop laser sintering printer" . More are certain to follow. Don't get overly excited just yet, though. To use the printer your desk has to be in a ventilated area near a source of running water [13]!

With regard to safety, the Nanosafety Research Centre of the Finnish Institute of Occupational Health, in collaboration with Aalto University and the University of Helsinki, has launched a study of 3D printing work environments [14]. They state that not enough is known about the emissions of 3D printers in general and how hazardous these emissions are. They recommend good ventilation while printing, both at home and in the workplace.

Up to now, various filaments of thermoplastics, nylon, wood, ceramic and viscoelastic (memory foam) materials as well as powders of tantalum, titanium alloys, cobalt chrome alloys, stainless steel, gold, silver, Inconel alloy, aluminum metals , sand, glass, ceramics and concrete have been used, with the list growing rapidly.

The 3D printing technique has also been adapted for basic research in materials science. The Missouri University of Science and Technology is working on developing stronger, more durable materials for NASA. Combining additive manufacturing with conventional approaches to create materials, researchers have been able to make steel parts that are 10% stronger than steel that is machined [15].

Using a technique they call "diffusion driven layer-by-layer assembly", researchers in Japan have constructed graphene, a single, tightly packed layer of carbon atoms that are bonded together is a single hexagonal honeycomb lattice. In this form, at one atom thick, it is the lightest material known to man, several hundred times stronger than steel and an excellent conductor of heat and electricity. Layers of graphene stacked upon each other form graphite[16]. Usually formed by vapor deposition, an expensive and complex process, this version of 3D printing has the potential to create porous three-dimensional structures for applications in devices such as batteries and supercapacitors [17].

Researchers at Oak Ridge National Laboratory used an EBM system fuzing together successive layers of metal powder while manipulating the process to precisely manage the solidification on a microscopic scale. The goal is the eventual development of a capability to design materials for specific applications [18-19]. Hofmann et al have adapted 3D printing to introduce radial grading of mechanical and physical properties in metals for space applications [20].

Marcus Ritlant [21] has compiled videos [22] showing the functioning of various types of 3d printers. These are very informative and well worth viewing.

The size range of 3D printers is fascinating. While the units shown in Figures 1.11 and 1.12 above are more or less typical for industrial and home uses, the actual range of printer sizes varies from huge to tiny. Figure 1.13 shows a Sciaky Inc metal printer, capable of making objects 110"x110"x110" in size.

*Figure 1.13  EBAM Metal Printer (Courtesy Sciaky Inc.-www.sciaky.com)*

If portability is required, there is the 3D pocket printer invented by Steven Middleton [23], Figure 1.14. At the other end of the size spectrum is the LIX 3D pen, advertised as the smallest 3D printing pen in the world.[24]. The pen and some examples of its use are shown in Figures 1.15 and 1.16.

*Figure 1.14 -- 3D Pocket Printer (Courtesy Steven Midddleton)*

*Figure 1.15 -- The LIX PEN (Lix Press Kit)*

*Figure 1.16 -- LIX Pen Creations (LIX Press Kit)*

There it is in a nutshell – 3D printers from room size down through pocket size. In the next chapter, we will look at some of the things that have been done with them.

# Further Reading

In addition to the sources already cited, material is available to provide an overview of 3D print technology. We mention these with some trepidation. There exists a huge literature on the subject. With it comes a wide variation of quality and usefulness. A few are very good, a few are so sparse or so poorly written as to be almost useless and the bulk of the material falls in between these two extremes. A number of books exist, most written in the last few years. Despite a few wishful titles, there is no ultimate guide to 3D printing as yet. Each of the books have useful material and, if taken all together, give a good overview of the technology and a few useful tips on making 3D models. Individually they address different aspects – the history of 3D printing, building a 3D printer, different types of 3D printers, making stuff with a 3D printer, software/hardware guides (we use the term loosely) and so on.

The following citations are intended to make you aware of some of the publications available. Their usefulness and quality you will be able to judge for yourself. With regard to the books, we recommend reading user evaluations that are published on Amazon to see it they fit your needs. Buying all is an expensive proposition.

<u>Articles:</u>

- 3D Printing Industry, "The Free Beginner's Guide", http://3dprintingindustry,com/3d-printing-basics-free-beginners-guide/

- Popular Mechanics article: "How to Get Started: 3D Modeling and Printing", http://popularmechanics.com/technology/how-to/tips/how-to-get-started-3d-modeling.../

We recommend a subscription to the 3D Printing Industry, the Inside 3DP and the Fabbaloo newsletters (free) to keep up with developments in the field.

<u>Books:</u>

- Barnatt, Christopher (2013), 3D Printing: The Next Industrial Revolution, CreateSpace
- Budman, Isaac and Rotolo, Anthony (2013), The Book on 3D Printing, Amazon Kindle edition

- Chua, Chee Kai and Leong, Kah Fai (2014), 3D Printing and Additive Manufacturing: Principles and Applications, World Scientific Publishing, available from Amazon
- Clearbrook, Tim (2014), 3D Printing:The Ultimate Guide Book, Amazon Kindle
- Frauenfelder, Mark (2013), MAKE: Ultimate Guide to 3D Printing, Maker Media Inc.
- Gibson, Ian, Rosen, David W. and Stucher, Brent (2010), Rapid Prototyping to Direct Manufacturing Technologies, Springer-Verlag, NY
- Hoskins, Steve and Hoskins, Stephen (2014), 3D Printing for Artists, Designers and Makers: Technology Crossing Art and Industry, Bloomsbury Visual Arts, NY
- Mesco, Bertalan (2014), The Guide to The Future of Medicine: Technology AND The Human Touch, Webicina Kft
- Smyth, Clifford (2014), Functional Design for 3D Printing: Designing 3D Printed Things for Everyday Use, Create Space Independent Publishing Platform
- Thornburg, David, Thornburg, Norma and Armstrong, Sara (2014), The Invent to Learn Guide to 3D Printing in the Classroom: Recipes for Success, Construction Modern Knowledge Press

# References

1. Hausman, Kalahani Kirk and Horne, Richard (2014), 3D Printing for Dummies, Wiley, Hoboken, NJ
2. Crawford, Stephanie (2014), "How 3D Printing Works", http://computer.howstuff works.com/3-dprinting1.htm
3. http://computer.howstuffworks.com/stereolith.htm
4. Burns, Marshall (1993), Automated Fabrication: improving productivity in manufacturing, PTR Prentice Hall, July 1, 1993
5. http://en.wikipedia.org/wiki/3d_printing
6. http://rapidtoday.com/mcor.html; white paper, "How Paper-based 3D Printing Works, www.mcortechnologies.com
7. white paper, "Why True Colour 3D Printing Capability Matters", www.mcortechnologies.com
8. Ragan, Sean and Stultz, Matt (2014), "Plastics for 3D Printing", Chapter 8 in France, Anna Kaziunas (2014), Make: 3D Printing, MakerMedia, Sebastopol, CA
9. France, Anna Kaziunas (2014), Make: 3D Printing, MakerMedia, Sebastopol, CA
10. http://www.explainingthefuture.com/3dprinting.html
11. http://www.buy3dprinter.org/3dprintingtechnologies/fused-deposition-modeling
12. www.additivefashion.com/3d-printing-basics-materials
13. www.sintratec.com
14. http://www.aalto.fi/current/news/2015-01-004/
15. www.sciencedaily.com/releases/2013/09/130917153717.htm
16. http://www.graphenea.com/pages/graphene
17. Zou, Jianli and Kim, Franklin (2014), "Diffusion driven layer-by-layer assembly of graphene oxide nanosheets into porous three-dimensional macrostructures", Nature Communications, 5:5254.A summary is inwww.sciencedaily.com/releases/2014/10
18. www.sciencedaily.com/releases/2014/10/141015130641.htm
19. Molitch-Hou, Michael (2014), "Oak Ridge National Lab Controls 3D Printed Metal at the Microscale", http://3dprintingindustry.com/2014/10/16/oak-ridge-national-lab-controls-3d-printed-metal...
20. Hofman, Doulas, C, et al (2014), "Developing Gradient Metal Alloys Through Radial Deposition Additive Manufacturing", Sci. Rep. 4, 5357; DOI:10.1038/ srep05357

21. Ritland, Marcus (2014), 3D Printing with SketchUp (Kindle Locations 290-291). Packt Publishing. Kindle Edition
22. http:// www.denali3ddesign.com/ video-guide-to-3d-printing-technologies/
23. http://middletonsa.wix.com/portfolio
24. www.lixpen.com

# Chapter 2

## *Applications of 3D Printing*

## 2.0  Introduction

Every new invention is motivated by the desire to do something that was never done before or improve on currently existing ways to solve a problem. Since the 1990's, the applications for 3-D printing have literally exploded as size limitations and costs have dropped and the list of materials that can be used with this technology has expanded dramatically. The applications can be grouped into several broad categories:

- Industrial
- Space
- Housing
- Clothing
- Medical
- Consumer-oriented

Let's look as a few examples in each category.

## 2.1  Industrial Applications

One attraction of 3D printing for commercial applications is the ability to make complex 3D prototypes or finished products that are not easily manufactured by conventional means. At their present stage of development 3D printers cannot crank out large quantities of identical parts at costs as low as can be achieved through mass production. There are other features, however, which have attracted the attention of large manufacturers such as Airbus, Boeing, GE, Ford and Siemens among others.

There are other features of 3D printing that are appealing in situations where time and cost are important. Compared to conventional (subtractive) manufacturing methods there is less wasted material. Conventionally manufactured products are often transported long distances, even across continents before reaching their final destination. With 3D printing, production and assembly can be local. When unsold products are discontinued, they often wind up in landfills. With 3D printing they can be made as needed [1].

Rapid prototyping is still the main attraction of 3D printing for industrial applications. Slowly, that is changing. Today, it is estimated that about 28% of money spent on printing things is for the final product, as opposed to a prototype [2].

Sciaky, Inc. [3] has developed a metal printer using electron beam additive manufacturing technology (EBAM) for printing enormous metal prototypes. Shown in Figure 2.1, the machine has a build volume of 19x4x4 feet and can 3d print such metals as titanium, tantalum, stainless steel and Inconel. The technology is being used by Lockheed and Boeing for the construction of jet fighter parts.

An alternate approach to a huge printer is a series of industrial size printers which can produce components of an object which can then be assembled to make the whole, something larger than the capacity of an individual printer. RedEye, a Stratasys company, has a print facility that is fully automated, consists of 150 3D printers which can print objects on demand for its clients (Figure 2.2). Teaming up with Lockheed Martin Space Systems, the Lockheed – RedEye team 3D printed two fuel tanks for a satellite simulation project for NASA. Using FDM printers, two weeks were required to 3D print the fuel tanks and 240 hours to complete the final assembly. The bigger of the two tanks is 15 feet long. Ten different pieces were required for the large tank and six for the smaller one using a polycarbonate material [3].

Nozzles are relatively simple devices, specially shaped tubes through which hot gases flow (Figure 2.3a). All jet engines use nozzles to produce thrust, conduct exhaust gases out of the nozzle, and to set the mass flow rate through the engine. Nozzles come in various shapes and sizes depending on the mission of the aircraft. Rocket engines also use nozzles to accelerate hot exhaust to produce thrust.

GE Aviation is pulling 3D printing out of the laboratory and installing it in the world's first factory to use this technology in the manufacture of jet engine fuel nozzles. Scheduled to open in 2015 in Auburn, AL, the plant will use 3D printing to additively manufacture the interior of fuel nozzles for the LEAP jet engine. Each LEAP engine will have nearly 20 fuel nozzles (Figure 2.3b) produced in this manner.

*Figure 2.1 -- 3D Metal Printer (Courtesy Sciaky, Inc. – www.sciaky.com)*

*Figure 2.2 -- Commercial Print Facility (Courtesy Stratasys)*

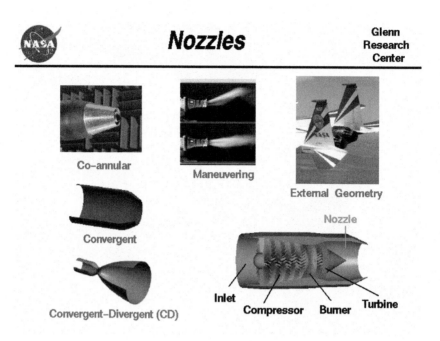

*Figure 2.3a -- Jet Engine Nozzles*

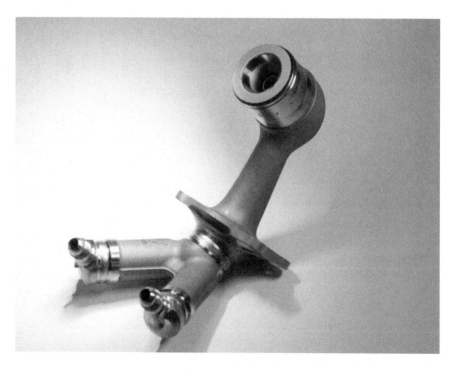

*Figure 2.3b -- Interior of Jet Engine Fuel Nozzle (Photo courtesy of GE Aviation)*

The LEAP fuel nozzles are 5 times more durable than previous models. 3D printing allowed GE Aviation engineers to design them as one part rather than the 20 individual parts required by conventional manufacturing techniques. Employing additive manufacturing also enabled engineers to redesign the complex internal structure required for this critical part, making it both lighter and more efficient. GE is also developing 3D-printed parts for the GE9X engine, the world's largest jet engine which will be installed in the next generation Boeing 777X long-haul passenger jet [4].

Ford [5-6] has employed stereolithography , laser sintering, as well as sand casting in developing rotor supports, transmission casings, damper housings and end covers for its C-Max and Fusion hybrids. This technology has also been applied to building four-cylinder ECOBoost engines, Ford Explorer brake rotors and F-150 exhaust manifolds.   Notable is the close relationship between Ford's research and development and its 3D manufacturing facility. CAD files are often sent back and forth between the two so that a design for a prototype can be built as a physical model, examined, tweaked as needed and then returned for fine tuning of the CAD model.

Engineers are constantly finding practical applications for 3D print technology. An example, reported in The Economist [7], deals with the physical principle that "...fluids flow more efficiently through rounded channels than they do around sharp corners, but it is very difficult to make such channels inside a solid metal structure by conventional means, whereas a 3D printer can do this easily. 3T RPD, a British firm [8] that offers additive-manufacturing services, printed a gearbox for a racing car with smooth internal pathways for hydraulic oil instead of drilled-out right-angle bends. The box not only allows faster gear changes but is some 30% lighter, says Ian Halliday, the firm's chief executive. A Boeing F-18 fighter contains a number of printed parts such as air ducts, for similar reasons."

## 2.2  3D Printing in Space

The National Aeronautics and Space Administration (NASA) has statically tested a 3D printed fuel injector for rocket engines.  Existing fuel injectors were made by traditional manufacturing methods. This required making 163 individual components and assembling them. Using 3D print technology, only 2 parts needed to be made, saving time and money as well as allowing engineers to build parts that enhance rocket engine performance and are less prone to failure [9].

The injector performed exceedingly well (Figure 2.4). Nicholas Case, the propulsion engineer leading the testing, summed up the case for including 3D print technology as part of the manufacture of rocket components: "Having an in-house additive manufacturing capability allows us to look at test data, modify parts or the test stand based on the data, implement changes quickly and get back to testing. This speeds up the whole design, development and testing process and allows us to try innovative designs with less risk and cost to projects." Figure 2.5 summarizes the advantages gained in using 3D printed injectors.

In September, 2014 NASA launched its first 3D printer into space [11]. Before the launch, it had to be tested and modified to work in a low-gravity environment [12]. Its short-term application will be for building tools for the International Space Station astronauts. In the longer term, 3D printers may be used to supplement the rations carried on space missions by printing food [13]. NASA is exploring ways to develop food that is safe, acceptable and nutritious for long missions. Current food systems don't meet the nutritional needs and 5-year shelf life required for a Mars mission. Because refrigeration and freezing require significant spacecraft resources, NASA is exploring alternatives. So far, some progress has been made on printing pizza [14 -15]. The pizza dough is easy enough to make, but at this stage you have to love it with ketchup and cream cheese!

*Figure 2.4 -- Test of 3D Printed Rocket Injectors (NASA/David Olive)*

## 3D Printing Application - Rocket components

NASA engineers 3D printed the engine parts for the Space Launch System (SLS), the vehicle slated to take human back to the moon.

Technology used: Selective Laser Melting (SLM)

An engine injector made,
**With conventional fabrication techniques of molding and welding:** In the range of US$250,000
**With 3D printing:** In the range of US$25,000 (reduced by a factor of 10)

Production times could also dwindle from six months to just weeks.

*Figure 2.5 -- Benefits of 3D Printed Fuel Injectors*

*Figure 2.6 -- Zero-Gravity Printer (Courtesy MADEINSPACE)*

Telescopes seem like simple devices but they are made up of many parts, are hard to build and hard to operate in space. Jason Budinoff of NASA Goddard is simplifying the process while working on the first space telescope made entirely of 3D printed parts.

Budinoff's design for the CubeSat satellite, Figure 2.7, is a fully functional 2-inch (50mm) camera. The only parts not printed are the glass mirrors and lenses. The remaining structures are printed from aluminum and titanium powders. The first prototype is strictly experimental and not yet ready for space flight. Further testing and development are in progress [16].

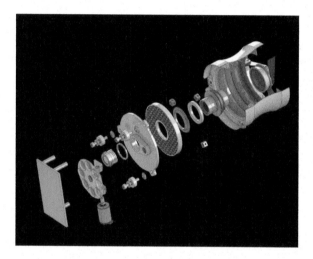

*Figure 2.7 -- 3D Printed Space Telescope (NASA Goddard/Jason Budinoff)*

## 2.3 Housing

3D printing has been used for some time to build architectural models. These help clients visualize the design, reduce the hours spent in crafting models and create a library of reusable designs. A few examples are shown below in Figures 2.8 – 2.11 . Thousands more can be found with a quick Google search. These models are not limited to a building here or a stadium there, but include scale models of cities. Figure 2.12 shows a scale model of Stockholm, Sweden, created by Mitek Gruppen with the help of a Stratasys Objet Eden 350V printer.

*Left: Figure 2.8 -- Paper-printed Highrise Building   (Courtesy Mcor Technologies)*
*Right: Figure 2.9 -- House  ( Courtesy Proto3000, www.proto3000.com)*

*Figure 2.10 -- Hanging Palm Lights (Courtesy 3DSystems)*

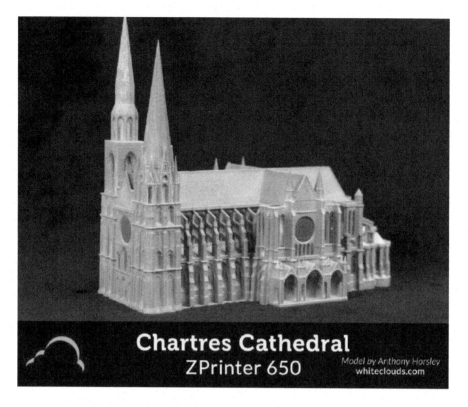

*Figure 2.11 -- Chartres Cathedral (Courtesy Whiteclouds)*

*Figure 2.12 -- Scale Model of Stockholm (Courtesy Mitek Gruppen/Stratasys)*

The question naturally arises: if one can build a model of a house, can one build a full-scale house? The short answer is – almost. There are house printers on the market. BetAbram, a Slovenian firm, is taking orders for the smallest of their concrete house printers (13x9.8x6.5 feet or 4x3x2 meters) [17] and is working on larger units. The procedure is to build one level of a house at a time. Once the first level is completed, the machine can be moved upward to build the next level and so on until the desired height is reached. It can be yours for a little over $15,000 (not including the concrete).

Billed as the world's first 3D printed house, the Canal House [18] in the Netherlands is under construction and is expected to be completed in 2015. It is being built at ¼ scale entirely from bioplastics. It is not expected to be an actual residence but a proof-of-concept undertaking. Each of the rooms will have furniture that illustrates the capabilities of 3D printing. A view of the house exterior is shown in Figure 2.13.

*Figure 2.13 -- The Canal House (Courtesy DUS Architects)*

A canal house is a symbol of Amsterdam. Its history goes back 400 years. Each house served several functions – trade, storage, living space, crafting. Each was richly ornamented and unique.

This Canal House serves as an exhibition, research and building site investigating the potential of 3D printing architecture. The project involves an international team collaboration in research and building techniques, thus linking science, design, construction and community.

The 3D printer used to build the Canal House, called the Kamermaker (Figure 2.14), is an upscale version of the Ultimaker 3D desktop printer [19]. The material used is a bioplastic made with 80% vegetable oil [20].

*Figure 2.14 -- The Kamermaker (Courtesy DUS Architects)*

As a brief aside, we should mention that people are looking into alternatives to concrete for building materials. A leading candidate is the humble soybean. Used both as food and as an ingredient in non-food products, students at Purdue University have now developed a soybean- based material which can be used for 3D printing [21]. Called Filasoy, it is a low-energy, low-temperature, renewable and recyclable filament created with a mixture of soy, tapioca root, corn starch and sugar cane. The aim is to provide an alternative to plastics for 3D printing as plastics are petroleum based and not derived from renewable resources.

Sofoklis Giannakoupoulous, a researcher at Barcelona's Institute for Advanced Architecture in Catalonia, is working on a soybean-based material that could be extruded to make structures that are more solid than concrete [22]. Still in the development stage, it is hoped that the material will be significantly cheaper than concrete and environmentally friendly as well.

Returning again to the real world where concrete is still the preferred material for construction, Andrey Rudenko, a Midwestern engineer, designed and built his own concrete printer and, following two years of research and development, has completed a small concrete castle [23]. The walls of the small fortress as well as the tower tops were fabricated separately and then assembled into a free-standing structure. Intermediate levels and the final structure are shown in Figures 2.15 – 2.18. He views this exercise as a proof-of-principle endeavor. His next effort will be a full-scale livable house [24].

*Figure 2.15 -- Layering Concrete for Castle (Courtesy Andrey Rudenko)*

*Figure 2.16 -- Intermediate Stage (Courtesy Andrey Rudenko)*

*Figure 2.17 -- Rounded Tower Base and Wall (Courtesy Andrey Rudenko)*

*Figure 2.18 -- Completed Castle  (Courtesy Andrey Rudenko)*

Ideally, 3D printing could be used to build entire full-scale houses or groups of houses. Prof. Behrockh Khoshnevis of the University of Southern California has developed a process called contour crafting which might make that possible. Contour crafting is a fabrication process by which large-scale parts can be fabricated quickly in a layer-by-layer fashion. It has the potential to build whole structures and substructures, complete with electrical, plumbing and air conditioning conduits embedded in the structure. Videos demonstrating the technique and its applications can be seen in [25].

## 2.4 Clothing

It's a nice, sunny day. You send the kids and the dog to the back yard to play. A few friends join them. You and the other moms have coffee while the kids play. They all come in when it's time for the friends to leave and you discover that yours have nearly destroyed their clothes. Your washer can deal with the grass and mud stains but the torn jeans…not so much. No problem. Their dimensions are recorded in the home computer. You call up the necessary data, input it into your 3D printer and by the time dinner is over they have new playclothes!

It's a nice dream, but we're not there yet. Real materials, such as silk and cotton, have an unfortunate tendency to burn during the printing process. The dream will have to wait a while. For now, 3D printing has attracted the attention of the fashion industry by way of a fashion-as-art concept [26]. The dresses consist of 3D printer fabricated components and the completed garment and its accessories are then finished by hand. A fascinating example of this approach is the Spire Dress designed by Alexis Walsh and Ross Leonardy [27]. It is made up of 400+ individual pieces, some in the form of spires, using nylon plastic. Then the individual pieces are assembled by hand to form the finished product, Figure 2.19.

*Figure 2.19 -- Spire Dress by Alexis Walsh and Ross Leonardy, Modeled by Jamie Simone (Courtesy Alexis Walsh)*

Two major drawbacks to printing full-size garments are the size of the printers and the ability to print natural materials. There are 2D printers which, given a t-shirt, can print virtually any imaginable design on it [28]. However, while there is a lot of interest, there is as yet no capability to print an entire t-shirt (or any other garment) from natural fabrics due to their unfortunate incendiary behavior. This is an area under study [28-29].

Shoe manufacturers have taken an interest in 3D printing. NIKE has used the technology for a football cleat [30]. A prototype for a lightweight plate attached to the shoe was made with a 3D printer. The company was able to manufacture the plates using 3D technology as well.

New Balance has done customization for runners as a pilot program to test the utility of 3D printing for athletic shoes. To make the shoe, New Balance fits the runner with a pair of shoes that used sensors to record data under simulated race conditions [31]. A video showing elements of the development process can be seen at [32]. Additional efforts in 3D printing shoes or shoe components are described in the article by Michael Fitzgerald [33].

The United States Army currently uses 2D programs to design clothing for soldiers. They are also investigating the feasibility of using 3D printing in the hope that it might eliminate or reduce the number of seams to make a garment [34]. Seams can cause discomfort in high heat and humidity, especially when a garment is worn with body armor. As a rule, the fewer seams, the greater the comfort.

## 2.5  Medical Applications

The healthcare sector has become a major user of 3D print technology. Figure 2.20 shows some of the applications. They range from creating customized crowns and braces for teeth, shells for hearing aids, various prosthetics and implantable devices, and models of various body organs to allow surgeons to refine their approaches and reduce the time needed for operations. Bioprinting, still in its infancy, will eventually allow customizing the delivery of medicines to specific organs, print human tissue and even cosmetics. Some recent articles [35 – 38] review aspects of the medical applications of 3D printers.

There is some speculation that the dental laboratory as we know it today may be replaced by 3D printing in the future. Traditionally, crowns are made in a dental lab. The dentist makes an imprint of the tooth (or teeth if a bridge is being designed) and sends it to a dental lab. Each lab technician, being human, has his or her unique style. This results in a variability in the final product from person to person and day to day, In addition, the traditional materials used in dentistry expand and shrink with exposure to temperature and moisture. This is difficult to control. The result of this on the patient is more time in the chair. A few days or a few weeks later the crown is sent to the dentist. Another visit is scheduled for the patient to fit the crown. Depending on the fit, a third visit may be required for a final adjustment.

It is now possible for a dentist to make a 3D scan of the tooth (or crown) and print it on the spot. Since it is made to measure, less time is required in the chair. As with almost all medical applications, the process is still in its infancy but shows great promise for increased patient comfort and (eventually) reduced cost [39].

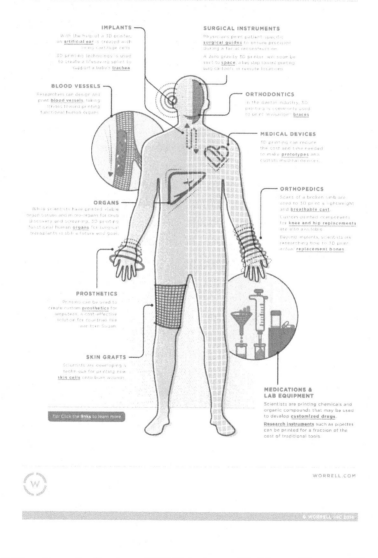

*Figure 2.20 -- Some Applications of 3D Printing to Healthcare (Courtesy Worrell, Inc.)*

Maxillofacial prosthetics (eyes, noses, ears, facial bones) are very laborious and expensive to produce. Ears and noses can cost up to $4,000 each. An impression is taken of the damaged area, the body part is then sculpted out of wax and that shape is cast in silicone.

Using 3D print technology [40 – 43], digital cameras are used to scan the injured area. A digital model is then created for the part, which incorporates the patient's skin tone. This information is sent to a 3D color printer. The cost of the printed part is about the same as that of a handcrafted prosthetic. The advantage lies in the fact that now a digital model exists. In the future, when replacements are needed for whatever reason, they can be made very cheaply.

3D printing has been effectively used to customize mechanical limbs. The usual goals are to add a capability missing due either to a birth defect or injury. Other reasons include improving the comfort and fit of an existing prosthetic device. As an example, consider a prosthetic hand designed for a man who, since birth, was missing a large part of his left hand. A high-tech prosthetic which cost over $40,000 was replaced with a 3D printed hand which provided him a stronger grip and cost much less, due in no small measure to the fact that it was developed by e-NABLE, a global online community of humanitarian volunteers which designs, builds and disseminates inexpensive functional 3D printed prosthetics [44-46]. The hand, called the Cyborg Beast, is shown in Figures 2.21 and 2.22. Many other examples of prosthetic devices, even one that serves as an exoskeleton permitting mobility can be found in the literature, as for example in [47] and [48].

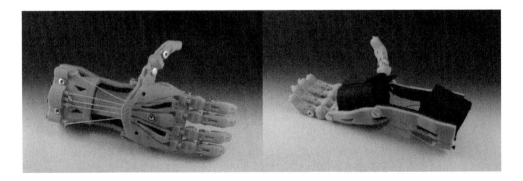

*Left: Figure 2.21 -- The Cyborg Beast Prosthetic Hand (Courtesy Prof. Frankie Flood)*
*Right: Figure 2.22 -- Reverse Side, Cyborg Beast Hand (Courtesy Prof. Frankie Flood)*

A six-week old baby's airways in one of his lungs collapsed causing him to stop breathing. Doctors at the C. S. Mott Children's Hospital feared that the baby would not leave the hospital alive unless something was done immediately. As luck would have it, the University of Michigan was developing a new bioresorbable device that could help the baby. The collaboration resulted in a custom-designed, custom-fabricated device, a splint, which was implanted in the airways to keep them open. The body will eventually absorb the device and the airway will stay open on its own [49].

Brain surgery requires drilling holes in skulls. Cranial plugs made on 3D printers can fill those holes. Cranial plates can replace large sections of a skull lost due to head trauma or cancer.

The Mayo Clinic and other hospitals have used 3D printed hip and knee replacements [50]. Dr. Barry Waldman of OrthoMaryland in Baltimore, MD used CT scans to cast an implant and manufacture the plastic cutting guides that direct the surgeon's incisions. The replacement joints were for a former athlete who had suffered for more than ten years with bowlegged legs bent six degrees out of alignment. The motivation for using 3D technology to create replacement joints was to minimize the amount of bone that had to be shaved off to install each implant. The article in The Baltimore Sun newspaper [51] quotes Dr. Waldman: "You can take an off-the-shelf implant and it may not fit. Plenty of times we'd be in the middle of a surgery and say : this is between a '5' and a '6' and we need a '5.5'. Well, there is no '5.5'".

3D printed replica models of body parts and organs have proven to be valuable in medical applications. They allow analysis of complexities and alternative approaches prior to a patient's surgery. This has had an impact of both the cost and duration of surgeries. The patient benefits if it is found that minimally invasive surgery is possible from shorter anaesthetic times and an earlier discharge [52]. At Houston Methodist Hospital, a stent graft made from a 3D image eliminates the need for open surgery for some patients suffering from abdominal aortic aneurysms [53].

An abdominal aortic aneurysm is an enlarged area in the lower part of the aorta. Since the aorta is the main supplier of blood to the rest of the body, a ruptured abdominal aortic aneurysm can be fatal. Surgery is required to treat this problem. This involves opening the belly and sometimes the chest to perform a complex operation.

In the new approach, a custom-made stent graft based on measurements from CT scans (Computerized Tomography – technique combining a series of X-ray views taken from many different angles and computer processing to create cross-sectional images of the bones and soft tissue inside the body) and measurements is placed on the aneurysm by going through the groin. Most patients go home after a few days with minimal pain or discomfort. The patients experience less blood loss. A shorter ICU (Intensive Care Unit) stay and a quicker return to a normal diet and regular activities than those who undergo an open procedure to fix this problem.

A number of bioprinters have appeared recently. According to [54], bioprinting is "using a specialized 3D printer to create human tissue. Instead of depositing liquid plastic or metal powder to build objects, the bioprinter deposits living cells layer by layer". The goal, still at least a decade away, is to build human tissues for surgical therapy and transplantation. Many laboratories are testing the concept by printing tissue for research and drug testing. The speculation is that patching damaged organs with strips of human tissue will occur in the near future [37].

## 2.6 Consumer-Oriented Products

Before the more than 250 3D printers now on the market became available, those interested in the subject had to build their own from scratch, known as DYI (Do It Yourself) or from kits. The only materials available then were easy-to-melt plastics. 3D printing then was the province of hobbyists interested in learning the new technique. The items printed were small, mono-colored and the end product of a learning process. With the greater selection of printers available today, a vastly increased selection of materials and greatly reduced costs, there now exist a very large number of

- Statuettes and figurines (bunnies, birds, cats, dogs, characters from movies or video animations, game pieces, etc.)
- Jewelry of every description, limited only by the artist's imagination
- Toys and action figures
- Art (practical things like vases to abstract art)
- Mugs
- Gadgets – phone cases are especially popular
- Household tools – screwdrivers, wrenches, broken part replacements
- Sunglasses of every description

and on and on and on. There is constant experimentation with and expansion of the capabilities of 3D printers. The main constraints are:

- The cost of the printer (fully assembled desktop units are available at prices ranging from under $400 to over $10,000)

- The cost of material (plastics for printing are not cheap. Don't be surprised if in the future the machines themselves will be available at giveaway prices because the profit will be in the materials they use)
- The size of the object to be printed (this can be overcome by printing individual parts and then assembling them info a finished whole)
- Constraints on printed objects by intellectual property laws. Remember, the lawyers haven't paid much attention to the 3D printing community yet, but that is bound to change.

Some examples of 3D printed objects are shown in Figures (2.23 – 2.2.6)

*Left: Figure 2.23 -- Bird House (Courtesy 3DSystems)*
*Right: Figure 2.24 Phone Case (Courtesy 3DSystems)*

*Left: Figure 2.25 -- Star Trek Figurines (Courtesy 3DSystems)*
*Right: Figure 2.26 -- Bug Droids (Courtesy 3DSystems)*

Food printing is being explored now that a larger selection of foods are available for that purpose [54]. Small and large food printers are available for specialized purposes. Some examples printed by the 3D Systems Chefjet printers are shown in Figures 2.27 – 2.30).

3D printing – it's not just for hobbyists anymore!

*Figure 2.27 -- Cake Decoration (Courtesy 3DSystems)*

*Left: Figure 2.28 -- 3D Printed Candy in Various Flavors (Courtesy 3DSystems)*
*Right: Figure 2. 29 -- 3D Printed Color Cake Topper (Courtesy 3DSystems)*

# References:

1. http://www.therepublic.com/article/10-industries-3d-printing-will-disrupt-or-decimate
2. http://www.theeconomist.com(node/21552892), "Making Things With a 3D Printer Changes the Rules of Manufacturing, 7/2/2014)
3. http://www.satellitetoday.com/telecom/2014/05/06/redeye-successfully-3-d-prints-two-fuel-tanks-for lockheed-martin/
4. http://www.gereports.com/post/91763815095
5. Murray, Charles (2013), "Ford Builds Metal Prototypes With 3D Printing", Design News, http://wwwdesignnews.com/document.asp?doc_id=256862
6. Bolton, Clint (2013), "Printing Out Barbies and Ford Cylinders", Business Technology, http://online.wsj.com/articles/SB1000142412788732337250457846956028212785
7. The Economist (2012 ), "Making Things With a 3D Printer Changes The Rules of Manufacturing", www.theeconomist.com/node/21552892
8. www.3trpd.co.uk/dmls.htm
9. www.space.com/topics/3d-printing/video entitled "3-D Printed Rocket Part Passes Biggest NASA Test Yet"
10. http://www.nasa,gov/press/2014/august/sparks-fly-as-nasa-pushes-the-limit-of-3-d-printing-technology
11. www.space.com/27008-3d-printer-launch-international-space-station.html
12. Snyder, M. P. et al (2013), "Effects of Microgravity on Extrusion Based Additive Manufacturing", AIAA Space 2013 Conference & Exposition, San Diego, CA, Sep. 10 – 12, 2013
13. www.nasa.gov/directorates/spacetech/home/feature_3d_food.html
14. www.space.com/topics/3d-printing/article: "How 3D PrintersCould Reinvent NASA Space Food"
15. www.space.com/21250-nasa-3d-foo-printer-pizza...
16. www.nasa.gov/content/goddard/nasa-engineer...
17. www.betabram.com
18. http://3dprintingindustry.com/2014/01/22/3dpi-tv-world's -first-3d-printed-house-begins-construction/
19. www.ultimaker.com
20. www.3dprintcanalhouse.com
21. http://www.3dprinterworld.com/article/purdue-students-invent-soybean-based-3d-printing-filament
22. http://textually.org/3DPrinting/cat_printing_a_house.html

23. http://www.designboom.com/technology/3d-printed-concrete-castle-minnesota-andrey-rudenko-08-28-2014/
24. www.totalkustom.com
25. http://www.contourcrafting.org
26. www.forbes.com/sites/rachelhennessey/2013/08/07/3-d-printed-clothes...]
27. www.newschool.edu
28. http://www.engineering.com/3D Printing/3D Printing Articles/Article ID/7114/The-Search-Begins-for-3D-Printed-Cotton-Textiles
29. http://www.fastcodedesign.com/3024883/forget-sewing...]
30. http://nikeinc.com/news/nike-debuts-first-ever-football-cleat-built-using-3d-printing-technology
31. http://www.boston.com/businessupdates/2013/03/08/new-balance-uses-printing: technique-customize-track-shoes/voGgY5NN9efZpCWrfqopTN/story.html
32. http://www.youtube.com/watch? V=qXfgihYUDhe&feature=youtube.be
33. http://www.sloanreview.mit.edu/article/with-3-d-printing-the-shoe-really-fits/...
34. http://www.army.mil/article/130211/Future_Soldiers_may_wear_3_D_printed_garments_gear
35. http://link.springer.com/article/10.1007/15548-010-0476-X
36. http://www.inside3dp.com/will-3d-print-medicine-within-10-years/
37. http://harvardsciencereview.com/2014/05/01/the-3d-bioprinting-revolution/
38. http://rt.com/usa/182120-3d-printer-drug-science/
39. www.stratasys.com, White Paper, "Perfecting Dental Treatment via 3D Printed Models and Removable Dies", 2014
40. http://www.huffingtonpost.com/2013/11/15/3d-printer-inventions_n_4262091.html
41. http://theguardian.com/artanddesign/architecture-design-blog/2013/nov/08/faces-3d...
42. http://www.replica.3dm.com/Article2.htm
43. http://www.replica.3dm.com/Article3.htm
44. www.enablingthefuture.org
45. http://enablingthefuture.org/upper-limb-prosthethics
46. http://enablingthefuture.org/prosthetics-meet-3d-printers-press-release
47. http://www.webmd.com/news/breaking-news/20140723/3d-printing
48. http://3d printingindustry.com/2014/02/22/walking-3d-printed-exoskeleton
49. http://www.uofmhealth.org/news/archive/201305/baby%E2%80%99A-life-saved-groundbreaking-3d-printed-device
50. http://www.webmd.com/news/breaking-news/2014723/3d-printing

51. Bodley, Michael (2014), "3D Printing Offers Twist on Conventional Knee Replacement", The Baltimore Sun, July 20, 2014
52. http://www.replica3d.com/medical/htm
53. Houston Methodist, "New Stent Graft made from a 3-D Image on the Patient's Anatomy. Option for People Suffering From an Abdominal Aortic Aneurysm", Science Daily, 5 Sep 2014, www.sciencedaily.com/releases/2014/09/140905152949.htm
54. Wong, Venesse (2014), "A Guide to All The Food That's Fit to 3D Print (So Far)", Bloomberg Business Week, Technology, http://www.business week.com/articles/2014-01-28/

# Chapter 3

## *Creating a Three-Dimensional Model: Scanning, Software and Hardware*

## 3.0  3D Scanning

3D scanners have been around since the 1970's. They have been used for a variety of applications, including surveying, terrain mapping, documenting construction and mining projects. Scans have been made of historical buildings and works of art, e.g., Michelangelo's David and the Pieta [1], to name but a few. Scans are regularly made of ships, consumer products, coins, medical devices and dental appliances, among many other items.

A 3D scanner creates a digital representation of a physical object. Therefore, if it exists and is accessible, it can be scanned. The data collected by the scanning process is called a point cloud. This is an intermediate step for creation of a mesh, also called a 3D model, a digital representation of the scanned object. The mesh can be used for:

- Visualization
- Animation
- Archival purposes
- Creating models for rapid prototyping or milling
- 3D printing
- Analysis of structures under a variety of internal or external forces using finite element or finite difference methods
- Computational fluid dynamics

After some additional processing, this is the information sent to a 3D printer which creates a physical object.

A mesh is a collection of polygonal shapes which approximate, to a desired degree of computational accuracy, a real world object. There are choices to be made in coming up with an acceptable mesh which is good enough for all practical purposes and still economically feasible. As a simple example, Figures 3.1 to 3.4 show meshes for a sphere. Figure 3.1 is a coarse representation using only 49 polygons. The surface of the sphere is approximated by straight line segments. As the number of polygons in the model increases, the representation looks more and more like its physical counterpart. This begs the question: how fine does the model need to be? That depends both on the accuracy required to represent the 3D physical model and the cost involved in printing the 3D model. Every time the number of polygons in a model is doubled, the time required to print goes up by a factor of 8. Thus the final model used for printing represents a compromise between accuracy and cost.

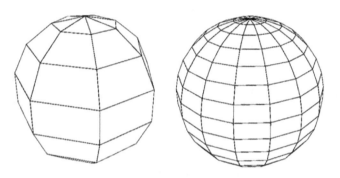

*Left: Figure 3.1 -- Sphere Model (49 Polygons)*
*Right: Figure 3.2 -- Sphere Model (225 Polygons)*

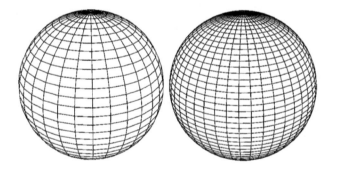

*Left: Figure 3.3 -- Sphere Model (900 Polygons)*
*Right: Figure 3.4 -- Sphere Model (2025 Polygons)*

For 3D printing, a solid model is the goal and the expected input for a 3D printer. The problem of mesh design and the compromises needed to get a sufficiently accurate mesh for an affordable computational price is a major consideration in computational solid mechanics [2-3] and fluid dynamics [4] where a mesh is just the starting point in studying the behavior of materials and structures under various external stimuli. If this topic interests you, consult the literature in these areas. In what follows, we touch on aspects of 3D scanning which lead to production of a 3D model.

## 3.0.1 Methods of Data Collection

Both contact and non-contact methods are used to collect information about 3D objects [5]. Depending on the nature of the object these can be used individually or combined.

In contact-based procedures, a probe touches various points of an object's surface to produce a data point (x, y, z coordinates of the location). Probes can be hand-held or part of a machine, referred to as a coordinate measurement machine or CMM. Such machines can be stationary or be in the form of portable arms. Figure 3.5 shows an example.

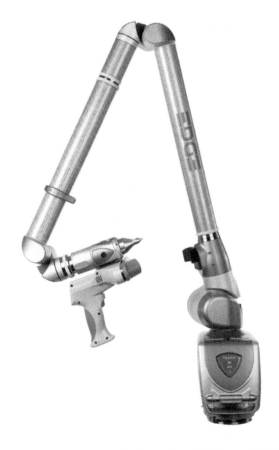

*Figure 3.5 -- FARO Edge Scan Arm (Courtesy FARO)*

Sometimes physical contact with an object is impossible, impractical or undesirable. Then it becomes necessary to resort to non-contact methods. These involve using lasers, ultrasound or CT machines. In laser scanning, a laser (red, white or blue depending on the application) passes over the surface of an object to record 3D information. As it strikes the object's surface, the laser illuminates the point of contact. A camera mounted in the laser scanner records the 3D distribution of the points in space. The more points recorded, the greater the accuracy. The greater the accuracy, the more time is required to complete the scan and the greater the complexity of the 3D model created from the data. With this method, it is possible to have very accurate data without ever touching the object. Examples of laser scanners are shown in Figures 3.6 and 3.7.

*Figure 3.6 -- FARO Laser Scanner (Courtesy FARO)*

Generally, contact digitization is more accurate in defining geometric forms than organic, free-form shapes. If it has been some time since you've taken an art class, *geometric* shapes have clearly defined edges typically achieved with tools. Crystals also fall into this category even though they are created by nature. Examples include spheres, squares, triangles, rectangles, tetrahedra, etc. *Organic* shapes are typically irregular and asymmetric. They have a natural look and a curving, flowing appearance. Organic shapes are associated with things from nature such as plants, animals, fruits, rivers, leaves, mountains, etc.

Laser scans produce good representations of an object's exterior but cannot record interior or covered surfaces. As a simplistic example, consider a hollow sphere. The laser scanner will accurately describe its general shape, but if there is anything inside the sphere, this would have to be detected with ultrasound or a CT scan.

The choice of which scanning technology to use will depend on the attributes of what you are attempting to scan, such as its shape, size and fragility. As a general rule, laser scanning is better for organic shapes. It is also used for high-volume work – scans of cars, planes, buildings, terrain, etc. It is the method of choice if an object cannot be touched, e.g., in documenting important artifacts. Digitizing is used in engineering projects where precise measurements of geometrically shaped objects are required.

Both methods can be combined when necessary.

The end result of a scan, regardless of which method is used, is called a point cloud. A 3D point cloud from a scan of the Coliseum in Rome is shown in Figure 3.8. The procedure for obtaining the point cloud and creating a 3D mesh can be found in [6-7].

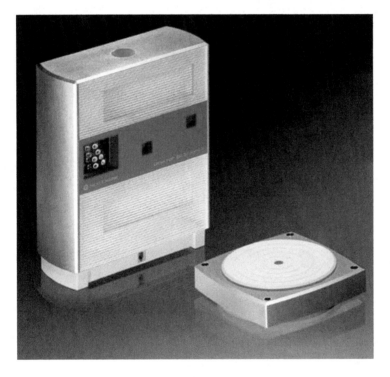

*Figure 3.7 -- NextEngine Desktop Scanner (Courtesy NextEngine, Inc.)*

## 3.0.2 From Point Cloud to 3D Model

The nice thing about point clouds is that they can be measured and dimensioned. This makes them valuable to architects and engineers. For the former, the ability to view and measure their project directly from their computer reduces the number of trips needed to the job site and thus reduces cost. For engineers, point clouds can be converted to surface models for visualization or animation and to 3D solid models for use in 3D printing, manufacturing and engineering analyses, including finite element analyses and computational fluid dynamics.

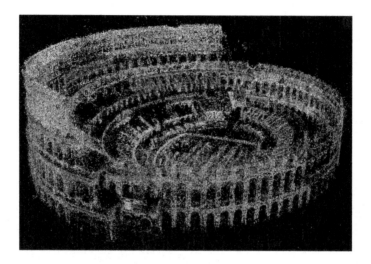

*Figure 3.8 -- A 3D Point Cloud (Courtesy Dr. Yasutaka Furukawa)*

Now that we have hundreds of thousands if not millions of data points obtained from 3D scans, we can go on to print our object, right?

Well, no.

*Figure 3.9 -- 3D Mesh Representation of the Coliseum*
*(Courtesy Dr. Yasutaka Furukawa)*

A hint as to why we need to do more is shown in Figure 3.9 The points have disappeared and have been replaced by a reasonable facsimile of a structure. There are two considerations before creating a 3D model:

- First, the point cloud data must be cleaned up a bit. A number of point clouds are generated in a scan to fully represent a 3D object. For purposes of analysis, these clouds must be merged into a single point cloud. This process is referred to as *registration*. Then some housekeeping must be performed on the consolidated cloud. Inaccuracies must be removed together with overlapping data, irrelevant points, noise, outliers, sampling error, missing data, misalignment, a varying sampling density, registration errors and other problems. Some of this will be done by software, some will require human intervention and manual correction. Keeping in mind, though, that laser scans can result in point clouds with 100,000 to over 1 million points, help from software will definitely be needed.

- Second, CAD software, which produces the 3D model suitable for various applications, doesn't know what to do with point clouds. Since CAD software expects to receive data in the form of surface representations of geometric forms and mathematical curves, the data must be translated into a form that it can interpret.

The software packages that create 3D models from point cloud data consist of both proprietary packages and commercially available ones. Among the popular commercial packages are (in no particular order):

- Geomagic [8]
- Polyworks [9]
- Rhino3D [10]
- Meshlab [11]
- Kscan3D [12]

and many, many others [13]. These software packages input the point cloud data, clean and organize it and produce 3D models in a wide variety of file formats, including the .STL file format popular in the 3D printing community. The final mesh can be achieved by way of a

- Polygonal mesh model
- NURBS (Non-Uniform B-Spline) model

Before continuing, we should point out that the business of going from point cloud data to a CAD mesh is a non-trivial exercise. There is no unique path for this. It is, in fact, an area of active research. A quick glance at some of the literature in this aspect of computational geometry will show this [14 -23].

With polygonal modelling one works primarily with faces, edges and vertices of an object. To make desired changes to a model, vertices can be repositioned, new edges inserted to establish additional rows of vertices and branching structures created. With polygon models the process is easier to grasp. However, as polygons are faceted, it can take quite a few of them to create a smooth surface. The more polygons, the greater the storage requirements. Even without this consideration, polygon modeling creates much larger files than NURBS modelling because the software keeps track of points and shapes in 3D space rather than mathematical formulas.

With NURBS modelling, one obtains smoother results. A NURBS object has only four sides. These are manipulated to create surfaces. This approach requires less storage than polygonal models. NURBS surfaces can be deformed, have shapes cut out of them, be stitched and blended together to form complex shapes. NURBS are constantly and always smooth as they are mathematically a continuous curve offering an easy way to keep smoothness within a model.

The requirements of your project – especially the time available for it - and the capability of the software package you are using – will guide your choice of the method you use to achieve the final mesh. In some cases, both will come into play. For low resolution polygon models, NURBS smoothing can be applied to provide a nice finish, polygon control and small file sizes.

## 3.1 Software for 3D Printing

In Chapter 1 we referred to 3D printing as a system (or a process). The key element in that process is the software. In 3D printing, as in other fields of engineering, the hardware development tends to precede the creation of software. Almost weekly there is an announcement of a new 3D printer either being developed or being brought to market. By contrast, some of the Computer-Aided-Design (CAD) software used to develop 3D models for those computers dates back to the 1970's. Software packages developed later tended to be evolutionary, rather than revolutionary, so that the philosophy that guided software developments some 40 years ago is still embedded in newer software, albeit with much improved user interfaces.

There are a number of reviews and listings [ 24 – 30] of software for developing 3D models. Here we follow the approach in [27] and categorize these as:

- CAD software
- Freeform modeling tools
- Sculpting tools
- Print Preparation and Slicing software

All will produce a 3D model suitable for printing. Each category, however, is aimed at a different audience. CAD programs typically deal with hard geometries and are well suited for engineering applications. Freeform and sculpting tools are aimed more at artists and creative modelers interested in animation, visual effects, simulation, rendering and modeling. Because 3D printers are very picky about the input they will accept, we need another software category that checks out the model and generates the g-code used by printers (preprocessing and slicing software).

## 3.1.1 CAD Software

There appears to be general agreement that, for beginners, the easiest to use programs are:

- 123D Design [31-32]
- 3DTin [33]
- SketchUp [34-36]
- TinkerCAD [37-39]

They have several characteristics in common. In most cases the basic software is free, with an option to purchase an advanced version as your capabilities and design needs increase. All are browser-based (typically current version of Google Chrome and Firefox). Tutorials are available to help get started. In the case of *123D* and *TinkerCAD*, these are extensive. All come with a library of primitive shapes (cones, spheres, squares, cylinders ,toruses) which can be imported to the work space and, by addition or subtraction, combined to form almost arbitrary shapes. *SketchUp* has extensive capabilities and documentation for applications to architecture and interior design in addition to civil and mechanical engineering. Most of the free versions and all of the advanced versions allow export of STL files to 3D printers, either your own or to a 3D print service.

The above are good for learning the basics of CAD software. There are several options available once you decide you need additional computing power. Some developers of 3D printers have proprietary 3D modeling software which is geared to their hardware and is not available unless you purchase their printer. The alternative is commercially available software, except as noted. These packages come with a price tag ranging from moderate to expensive. The learning curve rises steeply because of the added capability, and therefore complexity, of these packages. Most are intended for engineering applications. All come at minimum with tutorials on various aspects of the software. Training and consultation services are also available, at additional cost, for some of the software packages.

A partial list of software for intermediate, advanced and professional users includes:

- Solidworks [40-44]
- Inventor [45-48]
- Rhino3D [49-53]
- SketchUp Pro [54-56]
- CorelCAD 2015 [57-58]
- OpenSCAD [59-60]
- Free 3D CAD [61]
- PTC Creo Elements/ Direct Modeling Express [62]

As full descriptions of these software packages are available in the cited references, only a few comments need be made here. All but *Free 3D CAD* are available for purchase, the cost depending on the capabilities required. Most offer a free download of a downscale version to allow potential users to decide whether the program meets their needs.

*SolidWorks* and *Inventor* are comprehensive programs for engineering design and analysis. *SolidWorks* comes in three versions – Standard, Professional and Premium. There is a wide range of books, tutorials, guides, project files and videos to assist in various aspects of its use, from setup to engineering design applications. Training is available at additional cost. *Inventor* is available in two versions – Inventor and Inventor Professional. Extensive support is available ranging from help with installation, online tutorials and a community forum through 24-5 direct contact with and assistance from support staff.

*Rhino3D* most likely belongs in the next section dealing with software aimed at artists yet it is frequently used to develop models for 3D printing. Rhino is a NURBS program for creation, editing, analyzing and translating NURBS surfaces.

Like other programs in this category, it supports a variety of file formats. *Rhino3D* runs under Windows.

*CorelCAD 2015* has both 2D drafting and 3D design tools. It runs on both Mac and Windows computers. *OpenSCAD* is a 3D solid modelling program running under Linux/Unix, Windows and Mac OS X operating systems. Being a Unix-based system it is, in our opinion, not particularly easy to use unless one has first mastered Unix. *Free 3D CAD* is an open source, modular program that is designed as a parametric modeler – it allows modification of a design by going back into the model history and changing its parameters. The program is intended primarily for mechanical design. It is still in the early stages of development.

*PTC CREO Elements/Direct Modeling Express 6.0* is the free version of PTC Creo Elements/Direct Modeling 19.0. The program is aimed at engineers in search of a lightweight, flexible 3D CAD system who need to rapidly design and print one-of-a-kind objects and variations thereof. A variety of support services are available.

## 3.1.2 Freeform Software

CAD software is not the only path to the creation of a 3D digital model. For some time, 3D computer graphics software has been available offering features which permit 3D computer animation, modeling, simulation and rendering for games, film and motion graphics artists. The basic principles are taught at universities and the subject of many texts, among them the book by Vaughan [63]. Since the tools employed by graphic artists include polygons, NURBS and subdivisions, a 3D model suitable for printing is created in the process. As we haven't mentioned it before, a *subdivision surface* [64-65] is a method of representing a smooth surface using a piecewise linear polygonal mesh. A mesh is piecewise linear if every edge is a straight line and every surface is a plane. The most common examples are triangles in two dimensions and tetrahedra in three dimensions, though other options are possible. Given a mesh, it is refined by subdividing it, creating new meshes and vertices. The result is a finer mesh than the original one, containing more polygonal faces. This can be done over and over until the desired degree of surface refinement is achieved.

The software in this category includes:

- Autodesk Maya [66-69]
- 3ds MAX [70-75]
- Cinema4D [76-79]

- Blender [80-84]
- and others.

Both *Autodesk Maya* and *3ds Max* are the standard for the gaming and film industries. They used to be competing pieces of software, but are now owned by the same company. The only difference between these two is the layout and inclusion of certain tools. *3ds Max* works well with motion capture tools, while *Maya* allows you to import various plug-ins to create realistic effects. Many artists have a favorite between the two and will swear up and down by it. For the purposes of 3D printing, they are identical [84]. The full versions of both are free if you are a student. Otherwise, they are expensive.

*Cinema4D* is another 3D modeling, rendering, and animation tool. It has a large number of sculpting tools which let you mold the model as if it were clay to create shapes and contours. Unlike all of the other tools mentioned in this section *Cinema 4D* has a tool called the PolyPen Tool which lets you draw polygons on the screen, instead of starting with a base shape and working from there. This lets you create complex shapes very quickly and easily. For those who also work in gaming, it's designed to import seamlessly with the Unity 3D game engine, as well as auto-update in Unity when changes to a model are made in *Cinema4D*. Just like *Maya* and *3ds Max*, it is very expensive.

*Blender* is the free alternative to *Maya/3ds Max*. It's an open source program that aims to be just as good as its Autodesk cousins. Unlike the others, it comes with its own game engine and video editor included, which is useful for anyone looking to create a game on a budget. For 3D modeling, it's a good option to use if you aren't a student and want access to advanced software. If you need help getting your model from *Blender* to your 3D printer, Shapeways has a tutorial on how to export from *Blender* to a .stl file [81].

If you are trying to choose between the four, all of them have free trials (or in a case of Blender is just free). The best thing to do is play around with all of them and find which one works best for you.

## 3.1.3 Sculpting Software

Remember when you were young and played with modeling clay? You started with a ball of the stuff and then by pushing, pulling, pinching and squeezing you made a figure of some kind. Well, now you can do the same thing on a computer with the help of sculpting software. Make a model, export it as an STL file, send it to a 3D printer and you can relive the glory days of youth with greater accuracy and without getting your fingers dirty. What software lets you do this? Some examples are:

- 123D Sculpt [85]
- Cubify Sculpt [86]
- Leopoly [87]
- Sculptris [88]
- SculptGL [89]
- Zbrush [90-91]
- Mudbox [92]
- Geomagic Sculpt [93]

Digital sculpting is a relatively new but gaining in popularity. Many of the programs available use polygons to represent an object. Others use voxel-based geometry [94] in which the volume of the object is the basic element. Each has advantages and disadvantages for particular applications [95-96].

## 3.1.4 Almost But Not Quite

Let's assume that you have used one of the methods described above and have designed an object for 3D printing. Naturally, you are anxious to send it to send it to your own printer or to a 3D print service. First, though, there are a few things to consider:

- Your model exists in a mathematical world free of forces such as gravity. It is about to be transferred to the real world and printed with materials that have density and finite strength. Now is a good time to check the weight distribution. Otherwise your object may wobble or not be able to stand upright.
- Each printer bed has a finite size. If your object is bigger, you can always scale it down. Reality will intrude again if you do not ensure that critical dimensions such as wall thickness are of the minimum size required by the printer. If you are working with metals you might get away with thin elements but plastics are much weaker and far less forgiving, especially when heated. Even if you succeed in printing a thin member of your object, it may break during handling or shipment.
- Be careful of units. If your design is in millimeters, be sure that the printer does not expect centimeters or inches.
- To overcome the limitations of the printer bed or for artistic reasons, an object can be built out of dozens or hundreds of separate pieces. Hair,

buttons on a coat, different components of an object such as attire or accessories can be created as separate components in a 3D model. This won't work for 3D printing. Unless the individual parts are to be glued together after printing, the model received by the printer needs to be <u>a single seamless mesh</u>. Attaching a few parts to the object after printing is a nuisance. Attaching hundreds can be painful.

• By default, your object will print as a solid model. If this is what you want, fine. A solid model, however, requires significantly more material to print than a hollow one. Most printing services charge by volume. It is in your financial interest to print a hollow figure instead of a solid one if this is feasible. When hollowing a model, be aware of the minimum wall thickness that the printer you are using is capable of producing.

Other services (such as checking for water tightness and other geometric factors) are performed by print preparation software, or printer frontends. This type of software is a collection of utilities that check your 3D model and load STL files. The programs in this category [97] have an integrated slicing capability to create the layers in the z-direction and send the resulting G-code to the printer. Examples include:

- Repetier-Host [98]
- Printrun/Pronterface [99]
- MakerWare (for MakerBot printers) [100]
- Cura [101]
- ReplicatorG [102]

The slicing programs in frontend software also accept input from stand-alone slicing programs such as Slic3r [103] and KISSlicer [104].

Many tutorials [97] and articles [105-107] on preparing a 3D model for printing are available and should be reviewed prior to printing.

Finally you've reached the stage where you can make a printed object. But when it comes out of the printer or is returned by the print service, you see changes you'd like to make. Do you start the model creation process all over again? No. Software comes to the rescue again. Programs such as Netfabb [108], Autodesk Meshmixer [109] and Meshlab [110] allow you to edit your STL file and try again.

In Chapter 1 we referred to 3D printing as a system. We hope that you now see that software is the critical element in that process. New printers appear regularly.

Software development is an evolutionary and slow process. To paraphrase Commander Montgomery Scott (of the Starship *Enterprise*): Building 3D printers is easy. Designing comprehensive, user-friendly 3D modeling software – that's hard!

## 3.2 Hardware

Depending on the source that you consult, there are now between 200 – 300 3D printers on the market. These range from machines for industrial applications and manufacturing through specialized printers for medical research (bioprinters) and housing (concrete printers) down to consumer-oriented desktop-sized machines or smaller. Some of these have been developed by multi-billion dollar corporations and are focused on professional use. These market both the machines and products made with them and provide extensive support and training to their customer base. Many of the consumer-oriented machines have been developed by small companies (20 or fewer employees) who are very good at building 3D printers but are in no position to provide extensive support and training to their customers. A crystal ball is not required to conclude that sooner or later there will be a shakeout and it will most likely be in the consumer-oriented market. There simply is not that much demand for yet another 3D printed pink bunny or a figure of Yoda (Star Wars). Companies with ingenious designs but operating on a shoestring budget will be bought up by the corporate giants and incorporated into the fold. Many others, with only a shoestring budget, will fall by the wayside. Eventually, 3D printers will reach plug-and-play status just as 2D printers have, but it will not happen soon and it won't require 300 of them. Before that happens, though, 3D modeling software needs to be drastically improved for the non-engineer, non-artist market.

There are a number of reviews and evaluations of 3D printers [97, 111-118]. Almost all are focused on the hobbyist or home user. Of necessity, this limits consideration to FFF and SLA printers since currently these are the only ones safe for use in a home environment. Industrial additive manufacturing machines and materials are listed in the Senvol Database [119]. We will not repeat here material that is readily available on the internet. Instead, we point out some salient features involving the use of 3D printers.

## 3.2.1 Nice to Know

Before tackling the decision about which printer to get, if any, let's cover a little background, starting with

*Expectations:* There are a countable infinity of multicolored 3D printed images to be found on the internet. Some are incredibly lifelike and look like they've been printed with a variety of materials. In all probability, these were made with a high-end professional printer, then post-processed for hours or days by professionals. After looking at a number of these, it is only human to expect that when you buy your $500 home 3D printer you will be able to duplicate such results. Now is the time to remember a few facts: (1) SLA printers print in only one color; (2) an FFF multi-nozzle printer will give you some color but not the quality or resolution you get from a machine that costs 80 times more than your desktop; (3) you need a 3D model to even think of printing.

2D printing is far more advanced technologically than 3D printing, but even in 2D printing you can't turn on your equipment, press "PRINT" and expect a fully composed letter to come out. It won't happen unless you first use the word processing software to compose the letter. A similar situation exists in 3D. To get the printer to work, you first     need to learn CAD or any of the other software packages mentioned in Section 3.2. Learning any of these programs is time-consuming. After 2 weeks of study you will be ready to print out a simple design. After 2 or 3 years you will be able to do things that are astounding, not to mention getting a well-paying job as a software guru.

If you are not yet CAD-ready, you may download a file from the various services that provide 3D models. The number of such services grows regularly so, in theory, you should be in good shape there. You have no guarantee, though, that these files are all print-ready. Some may have errors, others designed for a printer that is not yours. If you can edit and fix them, fine. The rule that governs such purchases is still *Caveat Emptor* (Let the Buyer Beware).

*Materials:* Basically, FFF is the only really suitable technology for the home at this time. Unfortunately, current FFF printers produce an unappealing surface finish. If you need a nice finish, you also need chemicals to make your object look better.

Materials for 3D FFF printing are not cheap. If you are making small objects the cost is small. If you make large objects, the cost is naturally higher. If you plan to make things in bulk, the cost can be very high, so much so that it would behoove you to look into alternative methods of manufacture.

3D printers are good for awakening one's creativity. They allow tinkering with new designs and are good for learning about the technology. They are especially useful to making unique designs of objects that would be prohibitively costly and slower to make by other methods. This last feature is particularly appealing to someone who owns an older house that was built several years before plumbing and electrical standards were put in place. Even a small repair job, however simple and straightforward it may seem, can turn into a major hassle and involve a search of supply houses and specialty shops for parts no longer made. After that, a call to a plumber or electrician may still be needed. Two or three of those might easily pay for a low-end 3D printer!

Are the materials you will need for your project readily available and economical? Home printers work with a single material. If you plan to use multiple materials, you may need a more expensive machine. Are the materials available locally? If not, factor in shipping costs and time delays into your project.

Home printers have nozzles that clog, moving parts that break down. Can you repair the problems that will inevitably occur? If not, can you get support for your printer, either from the manufacturer, the retailer or locally? How long will service take – a day, a week, several months? If it breaks down and you need to ship it for repairs, who pays for the shipping?

Since you are printing layer by layer in the z-direction, the bonding will be imperfect so that, in effect, you are dealing with a laminated object. Laminates are inherently weaker than an equivalent object that is machined. Will your object have the necessary strength for its intended application? From a mechanical engineering standpoint, the greater the number of layers, the greater the degradation in its strength. If the object is to sit on a shelf, no problem. If it is to be put to some use, if bending is involved, will it be strong enough for all practical purposes?

In the next chapter we will consider factors that will help you decide whether or not to buy a printer. For now, there are two questions you must answer:

*What are you going to do with your printer?*

Having thought about and answered that, there is a second:

*How much time and effort are you prepared to expend mastering the software that allows you to create 3D models? How much money are you willing to invest?*

The two go hand in hand. Answer these, and all other considerations fall into place.

# References

1. http://graphics.stanford.edu/projects/mich, Levoy, M. et al, "The Digital Michelangelo Project: 3D Scanning of Large Statues"
2. Reddy, J. N. (2005), An Introduction to the Finite Element Method, McGraw-Hill, NY
3. Zukas, J. A. (2004), Introduction to Hydrocodes (Studies in Applied Mechanics #49), Elsevier, Amsterdam
4. Anderson, J., Jr. (1995), Computational Fluid Dynamics, McGraw-Hill, NY
5. http://www.dirdim.com/lm_everything.htm
6. http://dl.acm.org/citation.cfm?id=2001293
7. http://faculty,cs,tamu.edu/schaeffer/teaching/689 Fall2006/poissonrecon.pdf
8. www.geomagic.com
9. http://www.innovmetric.com
10. http://www.rhino3d.com
11. http://meshlab.sourceforge.net
12. http://www.kscan3d.com
13. http://www.simple3d.com
14. Fabro, R. (2003), "From Point Cloud to Surface: Intl. Archives of the Photogrammetry, Remote Sensing and Spatial Information Sciences, Vo.. XXXIV-5/W10
15. Patent US6253164B1, "Curves and surfaces modeling based on a cloud of points", http://www.google.com/patents/US6253164
16. Altman, M., "About Nonuniform Rational B-Splines – NURBS", http://web.cs.wpi. edu/~matt/courses/cs563/talks/nurbs.html
17. http://karol.hatzilias.com/medical-3d-scan
18. Mencl, R. (2001), "Reconstruction of Surfaces from Unorganized 3D Point Clouds", PhD Thesis, Dortmund University, GE
19. Tishchenko, I. (2010), "Surface Reconstruction from Point Clouds", Bachelor_Thesis, Swiss Federal Institute of Technology, Zurich
20. Kos.informatik.uni-osnabrueck.de/ICAR2013/#support, "Large-Scale 3D Point Cloud Processing Totorial – 16th Intl. Conf. on Advanced Robotics, Montevideo, Uruguay, 2013
21. http://seraphinacorazza.wordpress.com/2012/12/28/modeling-techniques-differences-between-nurbs-and-polygon-modelling-new-york-city-rooftop-3d-architecture-project-research/
22. Keller, P. et al (2012), "Surface Reconstruction Points from Unorganized 3D Point Clouds", Chapter 6 in Telea, A. C. (ed.), Reverse Engineering – Recent Advances and Applications, InTech. Available from

http://www.intechopen.com/books/reverse-engineering-recent-advances-and-applications/surface-reconstruction-from-unorganized-3d-point-clouds

23. Berger, M. et al (2014), "State of the Art in Surface Reconstruction from Point Clouds", http://lgg.epfl.ch/reconstar_data/reconstar_eg14.pdf
24. http://3dprinterhub.com/3d-printer-software
25. http://www.3dprintingpin.com/list-of-free-software-for-3d-printing
26. http://digits2widgets.com/3dprintingsoftware.html
27. http://3dprintingforbeginners,com/software-tools
28. http://explainingthe future.com/3d-printing-directory.html
29. www.3dprinter.net/directory/3d-modeling-software
30. http://www.3ders.org/3d-software/3d-software-list.html
31. www.123dapp.com/design
32. Cline, Lydia (2014), 3D Printing With Autodesk123D, TinkerCAD and MakerBot, McGraw-Hill/TAB Electronics
33. www.3dtin.com
34. www.sketchup.com
35. Ritland, Markus (2014), 3D Printing With Sketchup, Packt Publishing
36. Chopra, Aidan (2014), SketchUp 2014 for Dummies, Wiley
37. https://tinkercad.com
38. Kelly, James, F. (2014), 3D Modelling and Printing With TinkerCAD: Create and Print Your Own 3D Models, Que Publishing, sold by Amazon Digital Services
39. www.lynda.com – assorted video tutorials for TinkerCAD
40. www.solidworks.com
41. Planchard, David, C. (2014), SolidWorks 2015 Reference Guide, SDC Publishers
42. Jankowski, Greg and Doyle, Richard (2011), SolidWorks for Dummies, Wiley
43. Planchard, David, C. (2014), Engineering Design With SolidWorks 2015 and Video Instruction, SDC Publishers
44. Musto, Joseph and Howard, William (2014), Introduction to Solid Modeling Using SolidWorks 2014, McGraw-Hill
45. www.autodesk.com/products/inventor/overview
46. Hansen, Scott, L. (2014), Autodesk Inventor 2015: A Tutorial Introduction, SDC Publishers
47. Tickoo, Sham (2014), Autodesk Inventor 2015 for Designers, CADCIM Technologies
48. Shih, Randy (2014), Parametric Modeling With Autodesk Inventor, SDC Publishers
49. www.rhino3d.com
50. Cheng, Ron, K. C. (2014), Inside Rhinoceros 5, Cengage Learning, 4th ed.
51. (2012), Rhino 5 Training Bundle, Tutorial DVD, Infinite Skills

52. (2012), Learning Rhino 5 Training DVD, Tutorial Video, Infinite Skills

53. Buscaglia, Dana (2009), Rhino for Jewelry, lulu.com, 2nd ed.

54. www.sketchup.com/products/sketchup-pro

55. Gaspar, Joao (2013), SketchUp Pro 2013 Step-By-Step, GetPro Books, 1st ed.

56. Gaspar, Joao (2014), SketchUp Pro – New Features, GetPro Books, 1st ed.

57. www.coreldraw.com

58. Grabowski, Ralph (2014), Inside CorelCAD, 3rd ed.,
    http://www.coreldraw.com/us/product/inside-corelcad-ebook

59. www.openscad.org

60. Williams, Al (2014), OpenSCAD for 3D Printing, CreateSpace Independent
    Publishing Platform

61. www.freecadweb.org

62. www.ptc.com

63. Vaughan, William (2012), Digital Modeling, New Riders

64. http://en.wikipedia.org/wiki/Subdivision_surface

65. Reif, Ulrich and Peters, Jorg (2008), Subdivision Surfaces, Springer

66. www.autodesk.com/products/maya/overview

67. Watkins, Adam (2012), Getting Strted in 3D With Maya: Create a Project
    From Start to Finish – Model, Texture, Rig, Animate and Render, Focal Press

68. Derakhshani, Dariush (2014), Introducing Autodesk Maya 2015:Autodesk
    Official Press, Sybex

69. Palmer, Todd (2014), Mastering Autodesk Maya 2015:Autodesk Official Press,
    Sybex

70. www.autodesk.com/products/3ds-max/overview

71. Murdock, Kelly, L. (2013), Autodesk 3ds MAX 2014 Bible, Wiley

72. Murdock, Kelly, L. (2014), Autodesk 3ds MAX 2015 Complete Reference
    Guide, SDC Publications

73. Derakhshani, Dariush and Derakhshani, Randi, L. (2014), Autodesk 3ds MAX
    2015 Essentials: Autodesk Official Press, Sybex

74. www.maxon.net/products/cinema-4d-studio/who-should-use-it.html

75. https://www.udemy.com/blog/3ds-max-vs-maya/

76. Rizzo, Jen (2012), Cinema 4D Beginner's Guide, Packt Publishing

77. Kaminar, Aaron (2013), Instant Cinema 4D Starter, Packt Publishing

78. http://www.digitaltutors.com/software/b3qsICFQQQ7AodhzoAOg

79. http://www.maxon.net/en/products/cinema-4d-studio/who-should-use-it.html

80. http://www.blender.org/features

81. http://www.shapeways.com/tutorials/exporting-from-blender

82. http://www.blender.org/support/tutorials

83. Blain, John M. (2014), The Complete Guide to Blender Graphics, Second
    Edition: Computer Modeling and Animation, A. K. Peters/CRC Press, 2nd ed.

84. Simonds, Ben (2013), Blender Master Class: A Hands-On Guide to Modeling, Sculpting, Materials and Rendering, No Starch Press

85. 123dsculpt.com

86. Cubify.com/en/products/sculpt

87. www.leopoly.com

88. www.pixologic.com/sculptris

89. www.stephaneginier.com/sculptgl

90. www.pixologic.com

91. Spencer, Scott (2010), ZBrush Digital Sculpting Human Anatomy, Sybex

92. www.autodesk.com/Mudbox

93. www.geomagic.com/en/products/sculpt/overview

94. (1993), "Volume Graphics", IEEE Computer, v.26, #7, pp. 51-64, http://labs.cs.edu/labs/projects/volume/Papers/Voxel

95. De La Flor, Mike and Mongeon, Bridgette (2010), Digital Sculpting With Mudbox: Essential Tools and Techniques for Artists, Focal Press

96. http://en.wikipedia.org/wiki/Digital_sculpting

97. France, Anna, K. (2014), Make: 3D Printing: The Essential Guide to 3D Printing, Maker Media

98. http://repetier.com

99. http://reprap.org/wiki/printrun

100. http://makerbot.com/makerware/http://software/ultimaker.com/http://replicat.org

101. http://software.ultimaker.com

102. http://replicat.org

103. http://slic3r.org

104. http://kisslicer.com

105. http://www.shapeways.com

106. http://3d.about.com/od/Creating-3D-The-CG-Pipeline/ss/Preparing-a-Model-for-3D-Printing

107. http://images.autocad.com/adsk/files/tips-for-optimizing-your-model.pdf

108. 105. http://zoltanb.co.uk/tips-and-tricks-on-preparing-complex-models-for-ed-printing/

109. http://www.netfabb.com

110. http://www.meshmixer.com

111. http://meshlab.sourceforge.net

112. (2014 and 2015), "MAKE: Ultimate Guide to 3D Printing", Special Issue: 3D Printer Buyer's Guide, Maker Media, Inc.

113. (2014), "3D Printer Buyer's Guide for Professional and Product Applications", 3D Systems Corp., White Paper, www.3dsystems.com

114. Hoffman, Tony (2014), "The Best 3D Printers", http://www.pcmag.com/article2/0,2817,2470038,00asp

115. (2014), Inside3DP, "Top 10 Desktop 3D Printers", http://www.3dp.com/reviews/3d-printer-comparison/

116. Preet, Jessani (2014), "Chaos3D: Exploring the current state of the desktop 3D printer market", http://www.inside3dp.com/author/preet

117. Jackson, Bruce (2014), "10 Things You Need to Know Before You Buy a Printer", White Paper, 3D Printing Systems, Australia, www.3dprintingsystems.com

118. Evans, David (2014), "Where the 3D printing revolution falls short", http://www.inside3dp.com/author/evans

119. http://www.sculpteo.com/blog/2014/07/22/list-of-professional-3d-printers/

120. http://senvol.com/database/

# Chapter 4

## *3D Printing Without a 3D Printer*

So far we've talked about the different kinds of 3D printers, accessories, and different applications of 3D printing, now it's time to get into more detail on 3D printing itself.

## 4.0  To Buy or Not to Buy

First, the big question – should you buy a 3D printer yourself or use one of the many 3D printing services that are available? For some, the choice to buy or not to buy is a bit easier. Many companies can invest in a 3D printer (if not an entire line of 3D printers) because they have the capital to afford the printer(s) and know that they will be getting constant use out of the machine. The overhead cost is recouped in manpower savings by letting the printers run overnight. This allows them to create prototypes and finished products quickly. They can also recoup the cost of the printer(s) and supplies by working with other companies and individuals to offer 3D printing services. On the other hand, a hobbyist looking to purchase a 3D printer without that pool of funds needs to think more carefully about such a purchase.

For both a large company and the hobbyist there are several factors to consider when making the decision to buy a 3D printer. What kind of 3D printer do you want/need? How often are you going to use it? Also, how much are you willing and/or able to spend on the machine?

Let's start with the kind of printer. We've gone over the various different types of 3D printers in previous chapters, so you have an idea of what each kind of printer can do. You need to select the kind of printer you want based on what kind of things you are going to print. Are you making mostly household décor and gifts, other highly detailed finished products ready to be shipped when printed, or are you printing out prototypes of ideas before sending it off to get printed by a commercial strength 3D printer? These have different requirements for the kind of filament needed, the space of the print bed, and the print resolution. If you buy before you know what you want to print you might spend a lot of money to find out your printer can't do what you need it to do.

What kinds of things you want to print will determine what kind of printer you need. After that you need to look at how much you're willing to spend. Speaking of money, you can judge the quality of a 3D printer by its cost. If the price is too good to be true, it usually is. Printers under $600 often require assembly like IKEA furniture (except it's even less intuitive), have a small print space, and/or have low quality prints for various reasons. If you're trying to figure out how much a 3D printer is going to cost, assume an average of $1200 for an at home non-industrial strength printer. Add on the cost for filament, usually $30 per roll for ABS/PLA plastics, and you have a large start-up cost. When choosing a 3D printer you need to decide if having a smaller print space and limited color options (as smaller printers usually only have one print nozzle) is worth the saving a few hundred, or if you want to spend more for a larger machine with more options. Another financial consideration is determining what conditions your printed objects will encounter. 3D print services will have commercial strength printers that are far too expensive for most people to buy and can print in materials that can take a beating and keep on going. On the other hand, if you're printing out something that doesn't need to be super strong you can get by with the plastic filament that is used by less expensive printers.

Lastly, how often you intend to use the printer is important. If you will be turning out multiple prints on a regular basis, or need to be able to print on your own schedule, then the upfront cost is worth it. The convenience of not having to wait for your print to be shipped to you, the quick turnaround of designs, and the savings of being able to print your creations (as opposed to wasting expensive material when creating designs by traditional methods), makes having your own printer a sound investment.

If, on the other hand, you're new to 3D printing and looking to see what the buzz is about, wanting to print the perfect gift for someone, or doing any kind of work that will only occasionally use a 3D printer, then it's better to start off with any of the large number of 3D printing services available. There is no point in investing a few thousand dollars in machine and filament only to have it sit around collecting dust. Meanwhile, using a 3D print service will allow you to test out the quality of various machines and filaments with a much smaller up front cost.

In summary, if you plan to use a 3D printer frequently, as either part of your work or hobbies, and have a clear idea of what you want to do with your printer then it's worth the investment. If you only plan to use a 3D printer occasionally and/or are still figuring out what you want to do with the technology then we highly recommend using one of the many 3D printing services available.

# 4.1 3D Printing Services

If you want to get a model printed but don't have a 3D printer there are several places that will handle the printing for you. The time and cost to have your model printed will vary from place to place. Time is usually determined by how complex and how large the print is, as well as what other projects are in queue at the print location. Cost is also based on the size of the print, but just as important is the material the model is being printing in. Gold, stainless steel, and other metals will be more expensive then plastics. In determining price there is the base cost for printing, based on the material, and then an additional cost determined by the volume of the model. If you need more details on pricing, all of the services listed below allow you to upload your model and get a quick price quote.

A simple google search will give you several results for places that offer 3D printing services, with more being added all the time. Below is a list of some of the more popular 3D printing options. Some of these are designed for industrial use, while others are geared toward the hobbyist.

Shapeways [1]

Shapeways is a large 3D printing service. In addition to printing it has both a 3D model database and a service that connects you to 3D modelers you can hire to design a custom model to be printed. Shapeways uses industrial strength 3D printers to create high quality prints, shipped to your door. Each print goes through a checking, post production, finishing, and quality control phase before it is packed and shipped. There is a wide variety of materials available, including brass, bronze, gold, full color sandstone, and of course various kinds of plastics. They have also spent the past year and a half testing porcelain as a material and are currently working with experienced designers in a pilot program. If those tests succeed, 3D printed porcelain will be available to everyone.

Not sure what material you want? Shapeways can send you material kits so you can see and feel samples of what each material is like. There are three different kits that can be purchased and each ship in 3 business days. The basic kit costs $30 and includes $25 in store credit.

Most prints take 7 – 12 business days. Large prints, and prints using some metals, can take up to 20 business days to print.

Solid Concepts [2]

Solid Concepts focuses on printing for industrial prototypes and components. They have a wide variety of 3D printers including PolyJet, SLA, full 3D color machines, FDM, and cast urethane. The number of materials available varies for each printer type. Their website details the best applications for models created with each type of printer. In addition, Solid Concepts offers finishing services to smooth out the final product and remove the visible layers.

The time it takes to print a model is generally 2 – 5 business days, though complex prints may take longer.

3D Hubs [3]

3D Hubs connects people with 3D printers to people who want something 3D printed. Upload the file you want printed and you can get a quote from several different places that can print the file for you, based on your location. Each print location has reviews, just like Amazon, that rate the printer so you know what to expect. PLA and ABS filament are the most common print material; however depending on what locations are nearby you might be able to get a different material. Each print location also has a Hub Profile which details what 3D printer(s) they use, the cost, the print resolution, material and colors available for print, and the delivery time. Once an order is placed, both 3D Hubs and the print location is notified and a 3D Hubs representative is assigned to the order to ensure that everything goes smoothly.

i.Materialise [4]

i.Materialise is another 3D printing service geared toward the hobbyist. There are 17 different materials you can choose from, including ceramics, alumide (a combination of aluminum and polyamide powder), various kinds of resin, ABS plastics, and a rubber-like material. The site is simple and straight forward; simply upload your 3D model and select your material to get a quote and order. Like Shapeways, it also offers 3D models to purchase and a forum to connect with 3D modelers for a custom design. It has a forum section to discuss ideas with the i.materiaise community and get feedback, and a blog which posts sales and exciting things the company is up to.

The time it takes to print a model is generally 8 – 15 business days, though some materials can take up to 20 business days.

MakerBot [5]

If you happen to live in or near New York NY, Greenwich CT, or Boston MA then you can head over to one of the three MakerBot retail stores and have your model printed. Generally a model will take 24 hours to print. However, if the model is small and the store isn't busy, turnaround time can be as quick as 2 hours. While you're there you can step into the 3D Photo Booth and get a custom 3D model of your head and shoulders, and/or get an in-store demo from one of the staff. All of the models are printed using plastics.

The UPS Store [6]

Everyone is jumping on the 3D printing bandwagon, and that includes UPS. After a highly successful pilot program, UPS is now offering printing in 48 stores nationwide. The stores are all using the Stratasys uPrint SE Plus printer, which is a higher grade of 3D printer then the standard at home printers. While the service is geared toward small businesses, you can print anything from custom cell phone cases to architectural models. The time to print will take anywhere from 4 to 24 hours, depending on how complex the object you want to print is.

The list of participating stores can be found at the end of the chapter.

## 4.2 3D Model Repositories

So you know where you want to print your model – now you just need the model. There are several places online that will allow you to download 3D models that can be printed. From there you can either print the files on your own 3D printer, send the file off to be printed by someone else, or in some cases order the file to be printed from the same site that is offering it.

The major differences between the various repositories are the models they have available. If you can't find something you like from one location, look at another. Every site gives you the .stl file for the model, .stl being the universal "this is an object that is made to be 3D printed" file type. A .stl file can be read by all 3D printers, so once you have that there is no limitation on what kind of printer you can use.

We've already mentioned i.materalise and Shapeways as places that offer both 3D models and 3D printing services. Another newer site along the same lines is Pinshape [7]. They have a variety of highly detailed models, such as game miniatures and busts of fictional characters, so now you too can have your own bust of Severus Snape on your desk. While most of the models on Pinshape have a cost to download, there is a free section for those who are on a budget.

One of the most popular sites for finding 3D models is Thingiverse [8]. Operated by MakerBot, Thingiverse is one of the largest repositories of 3D models designed for printing. There is a wide range of models from the very simple to the very complex. You can download models designed for gaming, household décor, fashion, and other uses. If any of the models need to be printed in certain ways then instructions are provided. All of the models can be downloaded for free; however you either need a 3D printer or must send the files to a printing service to get the final printed model.

Other sites that also offer 3D models for download are (but not limited to):

- YouMagine [9]: Nearly identically to Thingiverse, YouMagine is run by Ultimaker.
- My Mini Factory [10]: This site has a combination of free and paid models available for download. Unlike other repositories they have a props and costumes section. This gives you access to 3D replicas of swords, guns, and various items from popular movies and games.
- Cults [11]: This French site also contains a mix of free and paid models. While it has a smaller selection then others, many of the models are very unique (such as a necklace that looks like a window plant) and are worth checking out.
- Autodesk 123D [12]: Another large repository like Thingiverse and YouMagine. All of the models on this site were made using various Autodesk 3D modeling software, such as 123D Design and Meshmixer.
- NIH 3D Print Exchange [13]: This site focuses on models for medical uses. The different categories of models available are "Medial and Anatomical", "Custom Labware", "Small Molecules", "Proteins, Macromolecules, and Viruses", and "Bacteria, Cell, Tissues, Organism." It also contains a small selection of video tutorials on how to use some types of 3D modeling software.

If the 3D model you select is designed correctly, downloading the .stl is all you need to do. However, you may need to clean up the model using software such as Meshmixer and Autodesk 123D. More information on how to use that kind of software will be covered in the next chapter. For now, all you need to know is that depending on how the model was designed and how large the print space of the 3D printer is, you may have to make some modifications. For example, you may need to cut the model in half so that it fits in the print space and then glue the pieces together once it has been printed. You may also need to add support structure to hold up pieces of the model that hang out from the rest [14-15].

## 4.3  3D Printing Considerations

When choosing which 3D print service to use there are several things you need to take into account; the time needed to print, the cost of printing, and the quality of the print.

First, the time needed to print. If you need your print done in a hurry that may limit what materials you can use, as some materials can be printed more quickly than others. In general, the larger the model the longer time you will need to account for. Secondly is cost. The higher the quality of the material being printed, the more expensive it will be. A plastic model will be less expensive than one made in bronze or gold.  Lastly, there is quality. Some materials are higher quality than others. Along with quality is choosing the right material. Some materials are better for household décor then they are for constant wear and tear. You may need to use a higher quality (and possibly more expensive) material if you need your print to be able to withstand some damage.

## 4.4  A 3D Printing Example

To put some of the information we've gone over so far in perspective, let's go over a practical example of using a 3D printing service. For this, we used a 3D model of a ratchet wrench NASA recently sent to their astronauts aboard the International Space Station. The wrench was designed by Noah Paul-Gin, approved by NASA, and then sent up to the specifically designed 3D printer which operates in a zero-gravity environment on the space station.  It's the first object to be 3D printed in space on request from one of the astronauts. The .stl file for the wrench is available as a free download on NASA's website [16-17].

After downloading the .stl file from the NASA website we uploaded the file to two different 3D printing services, Shapeways and i.materialise. Both sites asked if the model was created using inches or millimeters for measurements. This is done so that the website can accurately determine the dimensions of your model. We chose millimeters and next were given the option to select the material for the wrench and the make any size adjustments we might want. Along with the list of materials was the price of printing the model in each type of material.

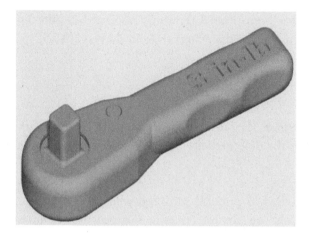

*Figure 4.1-- The 3D model of the ratchet wrench printed on the International Space Station.*
*(Image Credit: NASA)*

Shapeways also had another step in the process. Their site has a "3D Tool" menu that opens when you upload the model, and can be viewed again later if you want to make another print. This tool analyses the model and looks for possible printing problems. Some of the things it checks for are wall thickness, mesh integrity, part clearance, and detail work on the model (in this case the detail work is the engraved "3 in-lb" on the handle). It shows a 3D view of the model that you can rotate to see where any problem areas are. All problem areas are marked in yellow and you can click on the category that the issue is in to see why there is a problem. The tool gives you the option of having it automatically fix the problem areas for you. Once it generates the fixes it will show the new price that includes making the fixes to the model and the cost for printing in your chosen material.

In this case the tool found three small problems with the wall thickness on one of the moving parts on the back of the model. We chose the auto fix option and had it printed in "Blue Strong & Flexible Plastic Polished" material. The cost with shipping was roughly $27. Three days after the order was submitted, we received an email saying the model had passed manual inspection and was being sent to the printer. A week after that came another email saying the model had been shipped, and three days later it arrived. In total the process took two weeks. Below is an image of the final printed wrench.

*Figure 4.2 -- Print of NASA's racket wrench from Shapeways.*

We also tried to print the model with i.materialise. We say tried because a day after the order was submitted we were contacted by one of their Customer Support Engineers. She was worried that the moving parts didn't have enough clearance and that the model would be printed as one solid piece. She recommended that internal moving parts have a minimum of 0.8 mm clearance, and asked if we were willing to modify the model. We decided to cancel to order instead and see how the print from Shapeways came out.

Though the final print from Shapeways looks identical to NASA's final printed wrench, as feared the internal parts were printed solid. This could have been a problem with the model, or with the material that was chosen.

This is a good lesson in working with 3D printing – the first print will often not be perfect. You will need to make modifications after seeing the first result in order to refine the print and get your print to come out exactly as you want. This is a problem you will encounter if you use a printing service or own your own printer.

One of the authors (VZ) was fortunate enough to be given a tour of Northeastern University's 3D printing lab when we first started looking into the subject of 3D printing. One of the professors in charge showed her their bucket of failed prints. There were a number of reasons that prints wound up in the failed bucket. Sometimes the internal temperate of the printer was off. Other times the model had walls that were too thin and the support structure melded together with the model. One small model of Cinderella's Castle in Disney World had too fine a detail for its size, and so the area around the towers looked like a mess. Whether the fault lies with the machine or the design of the model, mistakes will happen. 3D printing is an iterative process. Be prepared to have a few messes when you start out.

## 4.5 Wrapping Up

Now you know where you can 3D print, where you can get models for printing, and you're starting to get an idea if investing in a 3D printer is right for you. However, what if you have the perfect idea for a 3D print that you just can't find online? In the next chapter we're going to take things a step further and teach you how to make your own simple 3D model, designed to work with any 3D printer.

# References

1. www.shapeways.com
2. www.solidconcepts.com
3. www.3dhubs.com
4. i.materialise.com
5. store.makerbot.com
6. www.theupsstore.com/small-business-solutions/Pages/3d-printing-locations.aspx
7. www.pinshape.com
8. www.thingiverse.com
9. www.youmagine.com
10. www.myminifactory.com
11. cults3d.com/en
12. http://www.123dapp.com/Gallery/content/all
13. http://3dprint.nih.gov/
14. www.hongkiat.com/blog/things-know-buying-3d-printer/
15. www.hongkiat.com/blog/download-free-stl-3d-models/
16. http://www.nasa.gov/mission_pages/station/research/news/3Dratchet_wrench/#.VJnpWMAAA
17. http://nasa3d.arc.nasa.gov/detail/wrench-mis

# Chapter 5

## *Create Your Own 3D Model*

Picking up from where we left off, let's assume that you either have a 3D printer or have chosen a print service to do the printing for you. Now you need to turn the idea you have in your head into a reality. To do that you need to know a few basic 3D modeling skills. We're going to show you how to make a simple 3D model using Meshmixer. Meshmixer is free software from Autodesk that allows you to combine, create, sculpt, and print 3D models.

## 5.0 Terminology and Advice

Before we get started, there are a few bits of terminology we should introduce. Throughout this chapter we will be talking about objects, planes, faces, and the grid.

An object is any three dimensional shape. Some models contain one object, while others are made from several objects combined together.

A plane is a flat surface. It can be an object or any kind of flat area.

The grid is the flat surface between the X-Y axes that is marked out by white lines. This is what you will be doing all of you work on and serves as a visual guide.

Faces are the various triangles that make up an object.

There are a few tips to keep in mind when working. You always want to use a regular mouse when working with Meshmixer. Most 3D modeling software is designed for a mouse instead of the trackpad that is in laptops.

Another useful tip is to save different versions of your project as you work. Keep a folder for each of your projects. Change the file name whenever you save your work. This means the folder will have lots of files named "ProjectName_version#". The reason to do this is to be able to easily go back if you have made a mistake, or to prevent losing your work if your computer starts to have issues and the file is lost or corrupted. This makes it easy to go back to an older version of the project instead of having to start from scratch.

## 5.1 Basic Movement and Object Manipulation

For this next section we will go over how to navigate within Meshmixer. It's recommended that you have the program open and follow along with the chapter. If you have any questions as we go, Autodesk has a series of video tutorials posted on YouTube that cover the basics of working in Meshmixer. These are a valuable resource and were very helpful to the authors when working on writing this chapter. A link to those videos can be found at the end of the chapter.

When you first open the program you'll see a few options. You can import a model you've started working on somewhere else, import a model from the 123D repository, open a Meshmixer file, or import one of three starting objects. Meshmixer doesn't let you start a new project with an empty grid, so click on "Import Bunny" for now while we show you the tools.

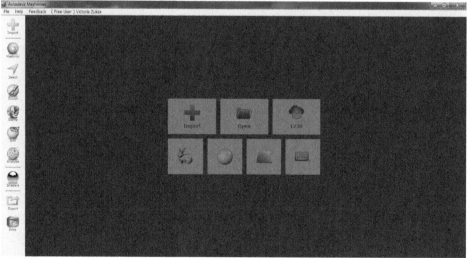

*Figure 5.1 -- Start-up screen for Meshmixer*

Once inside Meshmixer you will see a set of icons on the left, our bunny model in the 3D view on the right, and the menu bar on the top. If you hold down the space bar you can also make changes to some of the settings. The bunny is sitting on the grid, which is where you will do all of your work. Move your mouse to anywhere inside of the 3D view, where the bunny is, and hold down the right mouse button. As you hold down the button, move your mouse. You will see this lets you pan around, above, and below, the bunny. It's important to check all sides of your object while working.

Scroll the middle mouse wheel forward and backward. This lets you zoom in and out from the center of the screen. If you want to quickly center the camera on one part of the bunny, move the mouse over that part of the model and press C. The screen now moves so that that section of the object is in the center. You can then zoom in to get a closer look.

You can also move the grid around inside of the editing window. To do this, hold down the middle mouse wheel and drag the mouse. This combination of panning, zooming, and dragging allow you to move around your model in whatever way you need.

Holding down the mouse wheel will move the grid, but it won't move the object on the grid. To move the bunny, make sure it's selected. You can do this one of two ways.

Click anywhere on the bunny. It will be a white color when selected, and a grey color when not selected.

*Figure 5.2 -- Main screen inside Meshmixer*

Go up to the top menu and click on View, then Show Objects Browser. From there, click on "bunny.obj."

The Objects Browser is a list of all of the different objects in the scene. This is useful for selecting multiple objects at once, or if you have a large number of objects and want to make sure you are selecting the ones you want. You can select multiple objects at the same time using both of these methods. Select the first object, either by clicking on it or clicking on its name in the Objects Browser. Then press Shift and click on the second object, or its name. Both objects will now turn white.

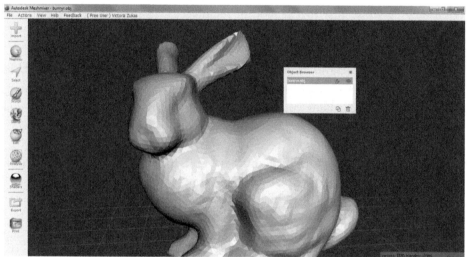

*Figure 5.3 -- The bunny object selected in the Object Browser*

In the Object Browser, you will see an eye shaped symbol on the same line as the name of our object. Clicking on that will hide an object. This is useful if you have several objects in the 3D view and only want to look at a few at a time. The center of the eye will be empty if the object is currently hidden. The object that you hide will disappear if there is more than one object on the grid. If there is only one object, like our current set up, or if the object that you hid is currently selected, it will be outlined in pink.

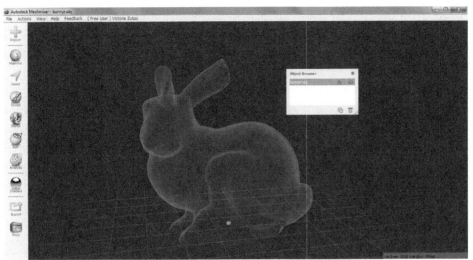

*Figure 5.4 -- Bunny object hidden in the 3D view.*

With the bunny selected, click the Edit icon on the left. This will bring up a new window. On that window select Transform. This brings up the manipulator on the bunny. The manipulator lets you move the object around, rotate, and scale it.

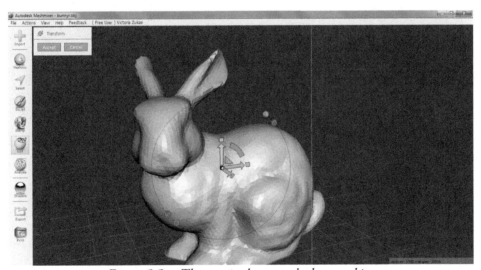

*Figure 5.5 -- The manipulator on the bunny object*

The long arrows show you the direction the model will move. Click and drag those to move the bunny along the X, Y, or Z axis. The colored triangles between the bases of the arrows let you move the object constrained to one axis. To do this, click and drag on the colored triangle.

At the tip of the colored arrows are matching colored boxes. These are used to scale the object. Clicking and dragging on one of those will let you stretch or squish the bunny along that axis. To uniformly scale the object, click and drag on the white box in the center of the manipulator.

Lastly there are colored half circles between the arrows. These are the rotation handles. Clicking and dragging on these let you rotate the object along that axis. When you start rotating you will see a number pop up telling you how many degrees you're rotating by. There is also a greyed out ring with ticks along it. Moving your cursor over the ring will snap the movement to the ticks. This let you make precise movements when rotating. You can change the increments of the tickets by pressing the up and down arrows.

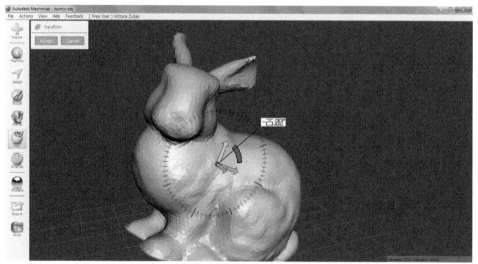

*Figure 5.6 -- The bunny rotated by -25 degrees.*

## 5.2 Editing and Creating Objects

We can now move around in the 3D view, move our object, and do basic manipulation. Let's take a look at the icons on the left side. Starting at the top, click on Meshmix. This opens a new window which lets you select various objects you can add to the 3D view.

At the top of the window you can select the category of objects you want to choose from. Once you find the object you want to add, click and drag it into the 3D view. There are two types of objects you can choose from – open parts and solids. Open parts have one side with no geometry. This is useful for adding parts onto another object. Solid objects are completely enclosed. Open parts have a blue and white circle icon in the lower right corner, and solid objects have a full blue icon in the same corner.

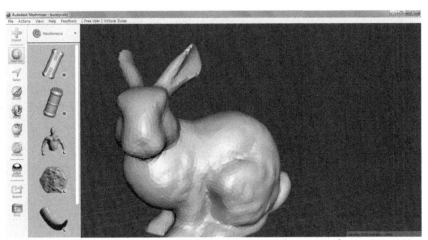

*Figure 5.7 -- A list of objects in the Meshmix window.*

Next is the Select menu. Often you will want to make changes to specific parts of the object. When you click on Select and mouse over the object you have selected, you will notice your cursor has a red dot in the middle and a grey circle around that. The area of the circle shows how large an area you will select on the object when you click. To select parts of the object, click and drag over the object. The selected sections will turn orange. If you make a mistake and select too much, mouse over the parts you don't want selected and hold down the Ctrl button while clicking and dragging over that section.

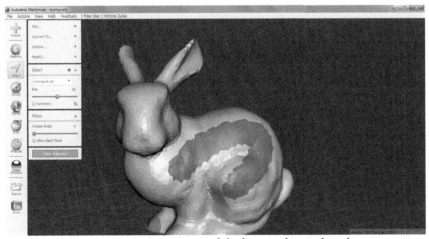

*Figure 5.8 -- A section of the bunny object selected*

In the Select menu is a Size option. This lets you tweak how large of an area you are selecting. When you're done with the area you're modifying, click the Clear Selection button at the bottom of the Select menu.

As you select an area on the bunny object you will notice the selection area has jagged edges. This is because you are selecting the faces that make up the object. You can see those faces by pressing the W key.

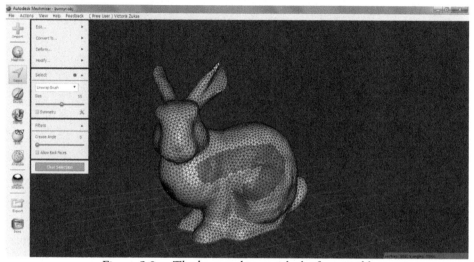

*Figure 5.9 -- The bunny object with the faces visible*

There is another method for selecting parts of the object. After clicking on Select, click anywhere in the 3D view that isn't the object. Then start dragging your mouse. You will notice a red line that follows your cursor. Drag around the part of the object that you want to select, and connect the circle you're creating to its starting point. Everything within that circle you drew will be selected.

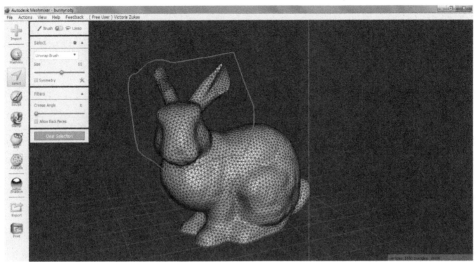

*Figure 5.10 -- Drawing a curve to select parts of the object*

Once you have part of the bunny object selected, new options will appear in the Select menu. The options you will most frequently use are Erase and Fill, Extract, and Separate under the Edit tool, and Transform and Soft Transform under the Deform tool. Erase and Fill removes the part of the object you selected and fills in any holes created by removing that part.

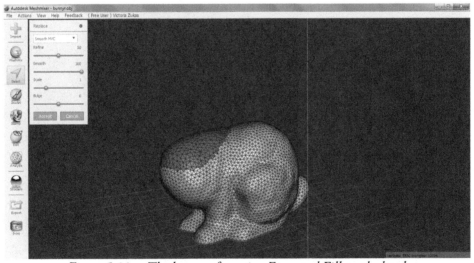

*Figure 5.11 -- The bunny after using Erase and Fill on the head.*

The Extract option creates duplicate geometry to the part that you selected, floating above the object. This new section can then be modified without modifying the rest of the object, in this case the bunny.

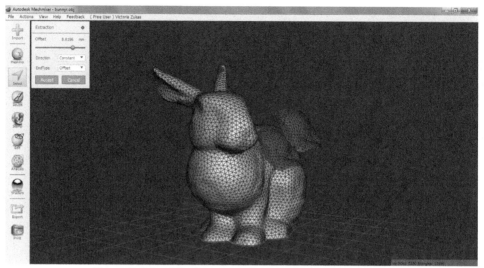

*Figure 5.12 -- The bunny with part of its side extracted.*

This new free floating section of geometry is still considered the same object as the object it was made from. If you want to change that, or to make any part of the object its own separate object, select what you want to separate and click on Separate. Now there are two object in our 3D View the bunny and the part of the bunny we separated.

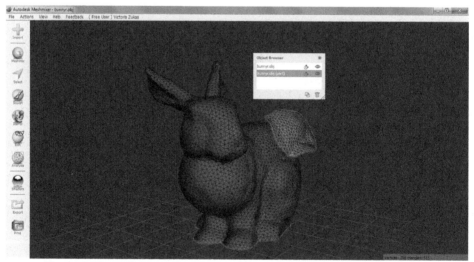

*Figure 5.13 -- The new object created from the extracted part of the bunny*

Under the Deform tool there is Transform and Soft Transform. Both of these options do the same thing, with an important distinction. Selecting either of these will bring up the manipulator from before. This functions exactly the same as when we were moving the entire bunny. When using the Transform tool the changes will be ridged, only modifying the area you selected.

Sometimes, such as when making an indent or a curved bend in an object, you want to affect the surrounding parts of the object as well. For this you would use Soft Transform. After selecting Soft Transform any part of the object that will be affected will be highlighted in orange. Move the Falloff slider on the Soft Transform menu to change how much of the object will be affected. From there, manipulate as you would before.

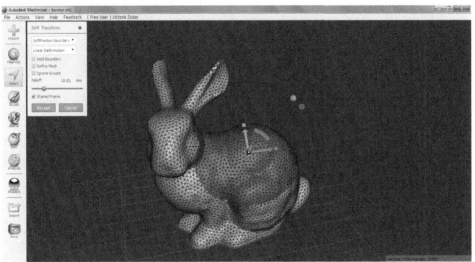

*Figure 5.14 -- The bunny with the Soft Transform tool active*

Lastly, there are a few options that don't appear unless you have multiple objects selected. To demonstrate this in the next few figures, a box has been dragged in from the Meshmix section. The box has been scaled down and moved so that it is popping out of the bunny's back.

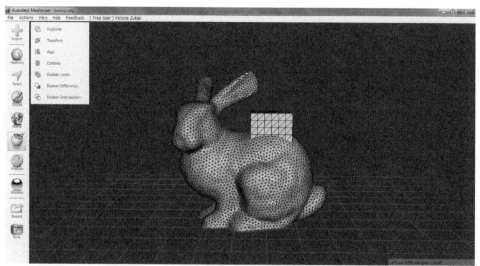

*Figure 5.15 -- The bunny and box object both selected in the 3D view*

If you select both the bunny and the box, you now have the options for Combine, Boolean Union, Boolean Difference, and Boolean Intersection. Combine makes Meshmixer consider both of these objects to be the same object, without changing them. Below you see the box is sitting inside the bunny as before, but only one object appears in the Objects Browser.

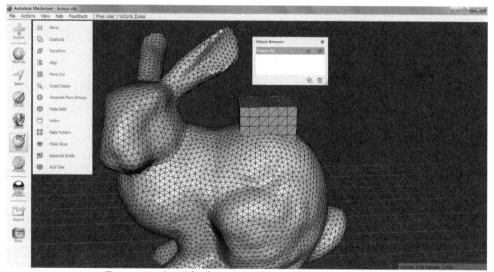

*Figure 5.16 -- The bunny and box object combined.*

Boolean Union merges the two objects. Unlike combine, it removes excess parts of the mesh. In this case, the bottom of the box is now missing.

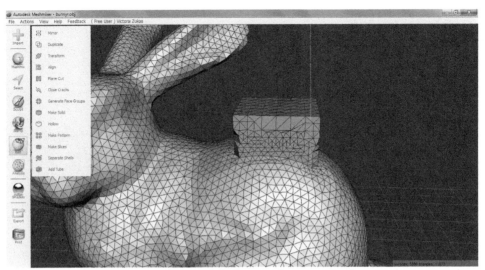

*Figure 5.17a -- The top of the bunny after using Boolean Union.*

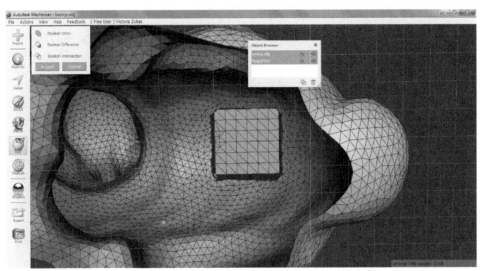

*Figure 5.17b -- The underside of the bunny after using Boolean Union.*

Boolean Difference removes part of one object where the two objects intersect. You can see the bunny now has a dent in its back from where the box and back part of the bunny met.

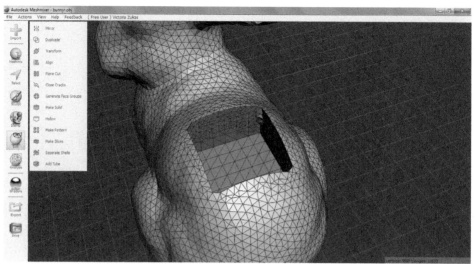

*Figure 5.18 -- The bunny after using Boolean Difference*

Boolean Intersection is the opposite of Boolean Difference. Instead of removing parts of the mesh where the two objects overlap, it removes everything except where the two objects intersect.

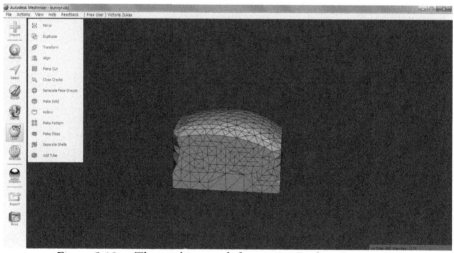

*Figure 5.19 -- The resulting mesh from using Boolean Intersection*
*on the bunny and the box*

Next is the Sculpt icon. As the name implies, these tools allow you to sculpt the model like clay. This tool has several advanced features. Since we will only be going over the basics here, we suggest playing around with the different options in this tool and seeing what kind of effects you can produce.

We move onto the Edit icon. The nice part about this section is that many of the names of the tools are self-explanatory. We've already gone over the Transform tool. Mirror does exactly what it sounds like it does; it creates a mirrored copy of the object. By default the new mesh created is mirrored along the Y axis. The manipulator will appear when you click on Mirror. Use the manipulator to adjust the axis the object is mirrored along and how much of the object is mirrored. When you are finished, click Accept. This new mesh will be fused with the original object. This is often used for making symmetrical objects. Just model half of the object and then mirror it to get the other side.

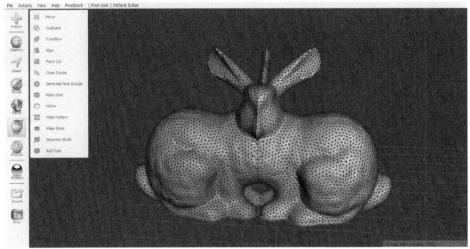

*Figure 5.20 -- The bunny after using the Mirror tool.*

Duplicate creates an exact copy of the object you selected. Unlike Mirror, the duplicated mesh is a different object, not attached to the original, and shows up as a new object in the Object Browser. The new object will appear directly over the old one, so it may not look like anything happened. Click on the Transform tool to move the new object.

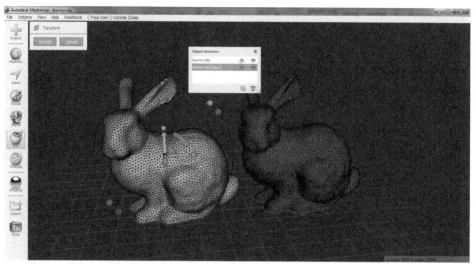

*Figure 5.21 -- The new object created by using the Duplicate tool on the bunny.*

Plane Cut allows you to easily delete parts of your object. When selected a plane that looks like the grid will appear. This grid can be rotated and moved in any direction. Part of your object will appear transparent and the rest will appear as normal. Move the plane so that the section you want to delete is transparent, and then click Accept. Now only the part of the object you wanted to keep remains. This is useful for quickly cutting an object in half.

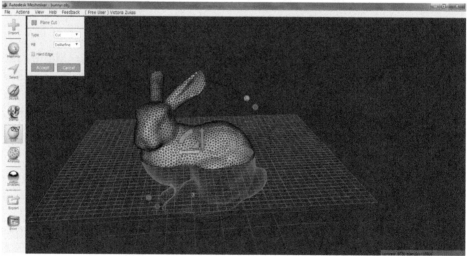

*Figure 5.22 -- The bunny while using the plane cut tool*

Generate Face Groups groups the faces that make up the object into different sections. This is useful if you want to be able to easily select and manipulate different portions of the object. Meshmixer looks at the object and determines where the groups will be placed based on the shape of the object. A widow appears with an Angle Thresh slider. By adjusting the value you change how many groups are created. The face groups appear as different colored zones on the object. To select an entire face group, double click anywhere inside of it.

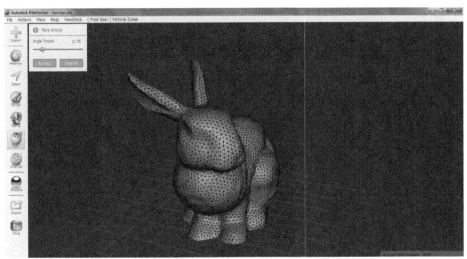

*Figure 5.23 -- The bunny object while generating face groups.*

Make Solid is used to prepare a model for 3D printing. This fixes certain problems in the object (such as overlapping geometry) that would cause issues when printing, and makes the model water tight. You can also set the minimum thickness of the walls in the object. If you know your printer needs your object to have a certain wall thickness, this is an easy way to make sure your object has that wall thickness. There are three settings in a drop down list at the top of the Make Solid window - Blocky, Fast, and Accurate. Blocky creates a pixelated version of your object. Fast is designed for fast printing. Accurate makes as accurate a solid object as it can. Similar to duplicate, the solid object is a copy of the original.

Other than the minimum wall thickness, you can also modify the Solid Accuracy and Mesh Density. Solid Accuracy determines how close to the original object the solid version is made. Mesh Density changes how many faces make up the object. Remember, the more faces that make up the mesh the more detail is has. Every time you make a change to any of the settings, click Update to have the changes appear on the object.

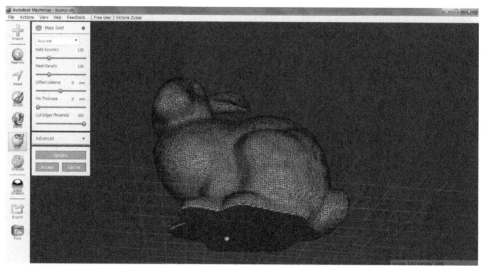

*Figure 5.24 -- The bunny after selecting the Accurate option in the Make Solid tool.*

Next we have the Analysis icon. Like the Make Solid option under Edit, these tools are used to prepare your model for 3D printing. The Inspector tool looks over your object and detects any problem areas. If we look at the bunny before using the Make Solid tool, you will notice the bottom is open. There is a blue sphere with a line pointing to where the issue is. Click on the sphere to fix the problem. In the case of multiple issues you can also click on Auto Repair All. There sphere and line will be blue, red, or purple depending on the issue the object has. When you are finished click on Done.

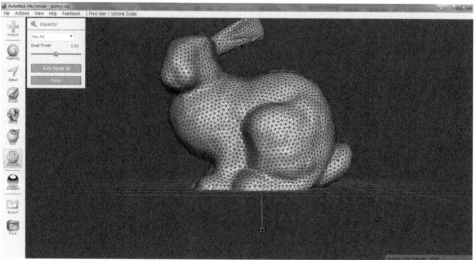

*Figure 5.25 -- The bunny inside of the Inspector tool.*

The Units/Dimensions option brings up a new window and a colored box around your object. This allows you to set the exact size of your object. The Overhangs tool is very important in preparing your model for printing. We mentioned in previous chapters that some models will need support structure when printing. This option allows you to create that support structure.

There are several settings in this tool. When you first select the Overhangs tool, it will outline in blue the sections of your model that need support. The support structure has three parts – the base, the post, and the tip. Click Generate Support to create support structure for your model using the default settings. Depending on the result you get, you may want to make some modifications to the settings and/or the generated result. To remove the current support structure click Remove Support. You should remove the current support structure before generating a new set.

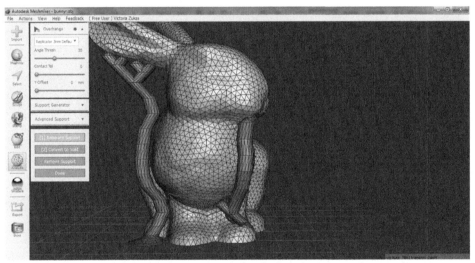

*Figure 5.26 -- Support structure for the bunny generated using the*
*Replicator 3mm default settings*

The Density option determines how much support structure is generated. The higher the density is set, the more struts will be generated and the more support the model will have. However, if this is set too high you may have a problem removing the structure after you print. You need to find the right balance between too little and too much support. The Post Diameter changes how thick the posts on the support structure are. The Tip Diameter changes how thick the tip of the post is. This is the part that directly connects to the model. This tool also requires finding the right balance for your print. If this setting is too low it won't do its job. If this setting is too high it will support the model very well and then be a pain to remove. The base diameter changes the size of the base.

Sometimes you may want to make minor modifications to the generated support structure. In the picture below the support structure was created using one of the default settings. Most of it is fine, but I would like to add some more struts to the bunny's chin, chest and left ear. There are several ways to do this. You can left click on the model (inside of the overhang area) or on the existing struts to auto generate another strut. This will only work on some areas on the model and struts. You can also left click on the model or strut and drag to create a new support strut. Drag mouse from where you clicked to where you want the support strut to end. If the new support strut you want to make will overlap with an existing strut or collide with the model a new strut won't be created. To force Meshmixer to make the new strut you want anyway, hold down shift while you drag.

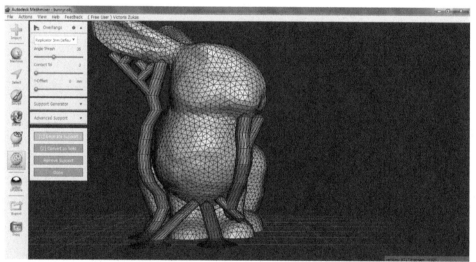

*Figure 5.27 -- The support structure created using the Replicator 3mm default settings, with a few more struts added.*

Last we have the Print icon. This is where you send your model to be 3D printed. The 3D view will change to show you the dimensions of the 3D printer you have selected. If you are printing at home there is a list of 3D printers to select from. You can also select to send your model to i.materalise, Sculpteo, or Shapeways for them to print for you.

The options available on this window slightly change depending on if you are printing from home or sending the model out to be printed. Once you have your printer or print service selected you can choose which material you would like to print your model in from the available drop down list. This menu appears below

Printer Properties if you are printing at home and below the name of the model if you chose a print service.

If your model isn't exactly where you want it on the print bed you can select the Transform button to move it around. Below the Transform button is the Repair option. If your model has any issues that you missed, it will have a button that says Repair Selected. This works the same way as the Auto Repair All button in the Inspector menu. Below that is the Overhangs option. If your model is missing support structure there will be a button that says Add Supports. This will generate support structure using the default settings for the printer you selected. If your model and support structure are ready for print, instead of the above mentioned buttons the Repair option will say Model Repaired and the Overhangs option will say Valid Supports. If you choose to print from a printing service, the Overhangs option will not appear.

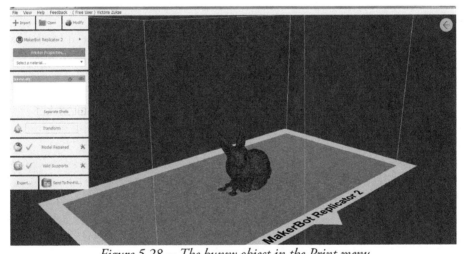

*Figure 5.28 -- The bunny object in the Print menu*

When you are ready to print click Send to PrinterName if printing from home or Review Cart and Order if printing from a printing service. If you are printing from a print service a cost estimate will be displayed so that you know how much your model will cost to print before you click Review Cart and Order. The price will change depending on the material that you selected.

If you selected Review Cart and Order, a new window will pop up with your shopping cart. This functions the same way as any online store. Click Order and the website for your selected printing service will open with your model already uploaded to their site. From here on you complete the purchase of your model through that service's website.

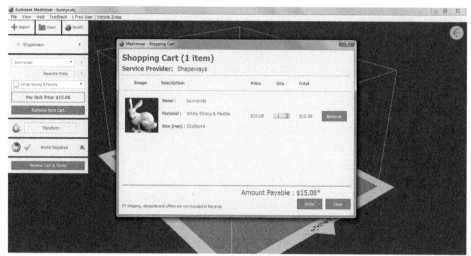

*Figure 5.29 -- The shopping cart window that appears after clicking Review Cart &*
*Order*

Now you know the basics for making a model and sending it to print in Meshmixer. Next, we'll walk you through a practical application of what we just went over. This section will walk you step by step through making your first 3D model.

## 5.3  Making Your First 3D Model

For this exercise we're going to walk you through how to model a flathead screwdriver. This tutorial assumes you are using Meshmixer 2.8, which at the time of writing is the most recent version of the software. Start with a new Meshmixer file. On the starting screen select the sphere to import. We will be deleting this object in a moment, but we need to start with something on the grid.

There are three parts of a screwdriver; the handle, the shank (the long cylinder attached to the handle) and the blade at the end of the shank. We'll start with creating the handle. Click on Meshmix, and then select Primitives from the group at the top of the Meshmix window. You will see a list of basic objects. About five shapes down is the cylinder.

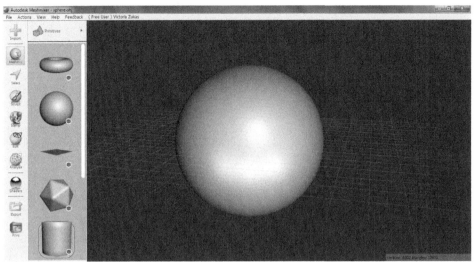

*Figure 5.30 -- The Meshmix window open with the cylinder outlined.*

Click on the cylinder and drag it into the 3D View. Next, click on the blue rotation handle and rotate the cylinder 90 degrees.

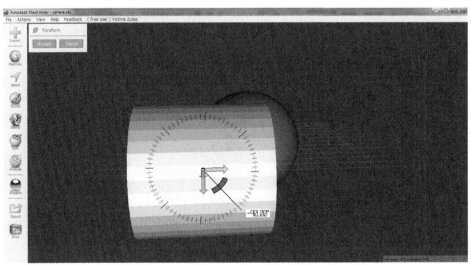

*Figure 5.31 -- Rotating the cylinder 90 degrees.*

Click on the green box above the green arrow on the manipulator to stretch out the cylinder. From here on it will be useful to see the faces that make up the object, so press the W key to make those visible. You will notice the faces divide our cylinder into 6 even sections. When your cylinder looks like the picture below, click Accept.

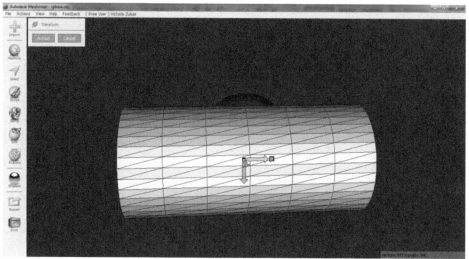

*Figure 5.32 -- The cylinder after scaling along the Y axis*

We no long need the sphere we imported in the beginning, so click on that and hit the delete key. Select the cylinder again. Click on the Select icon. Next, select the second through fourth sections, counting from the left. As you select these parts of the object they will turn orange. Don't forget to pan around your model and check to make sure you didn't select any extra faces by mistake.

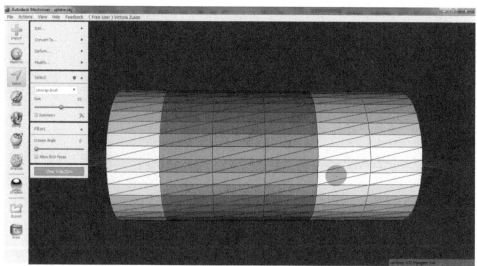

*Figure 5.33 -- Sections two through four of the cylinder selected.*

For this part we're going to make an indent in the handle. Still in the Select menu, go to Deform, then Soft Transform. The entire object will turn orange. Since we only want to affect the three sections we first selected, move the Falloff slider to the left until only the three sections we first selected are orange.

On the manipulator there are two circles in the upper right, one with an L and one with a W. The one that is currently selected will be green. If L (for local) is selected, the object will be transformed relative to the position it's currently in. For example, if the object is tilted moving it to the left will also move it down and to the left. If W (for world) is selected, then it will move in the direction indicated to matter which way the object is facing.

Select W on the manipulator, then click and drag to the left on the center white box. This will scale that section of the cylinder inward.

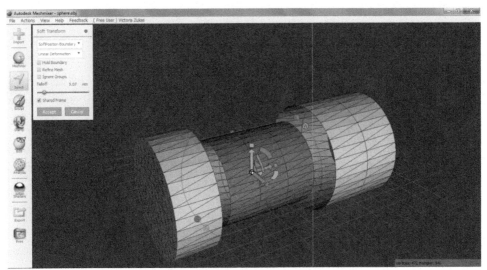

*Figure 5.34 -- Scaling the selected sections inward*

When the sections you have selected look like the picture above, click Accept. This will create the indent seen in the picture below. Feel free to adjust how much you scale this section to create more or less of an indent if you wish.

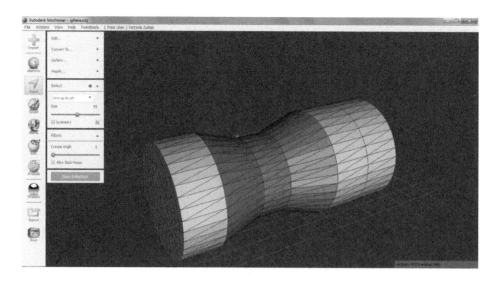

*Figure 5.35 -- The handle with the newly created indent*

Click on Clear Selection to deselect the indented section. Click on the Select tool again and select the top of the first section of the cylinder. You only want the flat circle at the base, not the rest of the section.

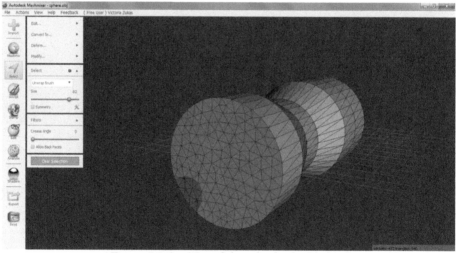

*Figure 5.36 -- Top of the cylinder highlighted*

Again, go to Deform and Soft Transform. This time we don't have to make any adjustments to the Fallout. Click on the blue arrow on the manipulator and move it to the right to squish this section towards the other side. When it looks like the picture below, click Accept and Clear Selection. Now the front part of the handle is finished.

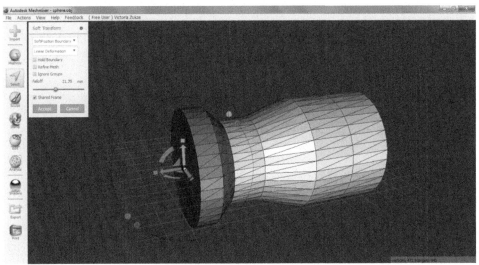

*Figure 5.37 -- Finished look of the front part of the handle*

Select the bottom of the cylinder. Go to Deform and then Transform. Drag the blue arrow to the right to extend the back end of the handle. Now our handle has the proper proportions. You will notice the new part we pulled out is green. This happened because Meshmixer automatically created a new face group for this part of the object.

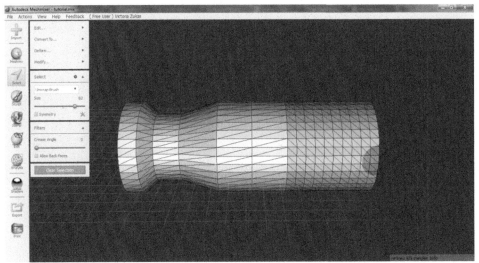

*Figure 5.38 -- New faces added to the end of the handle*

This next step is for ascetics. We're going to round off the end of the handle. The bottom of the cylinder should still be selected. Pull it towards the right using the Transform tool just far enough to make one new section. This new section should be the same size as one of the sections in green. Then, click on the middle white square in the Transform tool and drag the mouse to the left. This will pull the edges inward. The new section you just manipulated will look like the picture below.

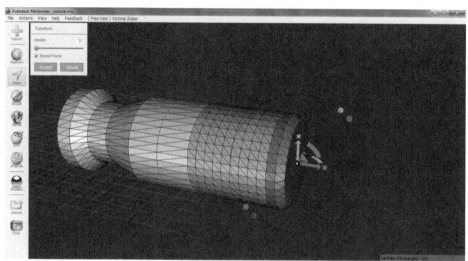

*Figure 5.39 -- The handle after creating a new section and pulling it inward*

Click Accept. This will make the Transform tool go away, so click on Edit, Deform, and Transform to bring it back. Now repeat the above steps three times (Make a new section, use the middle square on the Transform to pull it inwards, and Accept the change.) Make sure you don't skip selecting Accept each time. Your handle should now look like the picture below.

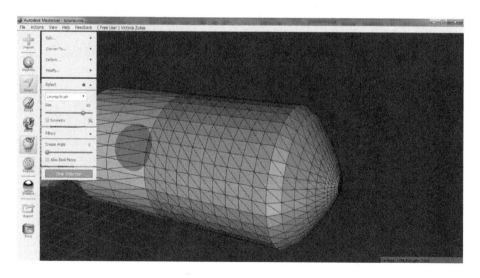

*Figure 5.40 -- The handle with a rounded off end*

Now we're going to add detail to the handle. For this next part to work we first need to make it a solid object. Clear your current selection and then select the handle. From there go to Edit and then Make Solid. Under the Make Solid window select Accurate from the top drop down list. Set the Solid Accuracy to 150, the Mesh Density to 80, and the Min Thickness to 5 mm. After you make those two changes click Update and then Accept.

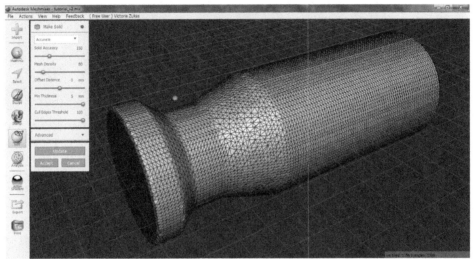

*Figure 5.41 -- The handle after using the Make Solid tool*

Most screwdrivers have indents along the length to allow for an easy grip. We're going to make those indents. In these next few steps we're going to create several cylinders and use the Boolean Difference tool to carve out the indents in the handle.

Go back to the Meshmix icon and drag another cylinder into the screen. Rotate it 90 degrees, as you did for the handle. Then use the manipulator tool to shrink its size and stretch it out until it looks like the picture below.

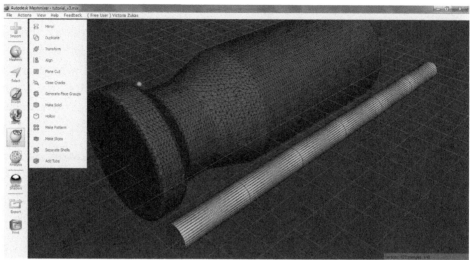

*Figure 5.42 -- The new cylinder added to the 3D view*

When it looks right click Accept. In the same menu click on Transform and move the cylinder so that it intersects with the back and front part of the handle but doesn't touch the indent near the front.

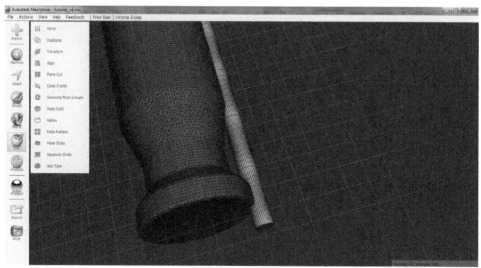

*Figure 5.43 -- Correct placement of the new cylinder*

Still in the Edit menu, click on Duplicate. Select transform, and move the new copy of the cylinder to the opposite side of the handle. It should intersect with the handle the same as its pair on the other side.

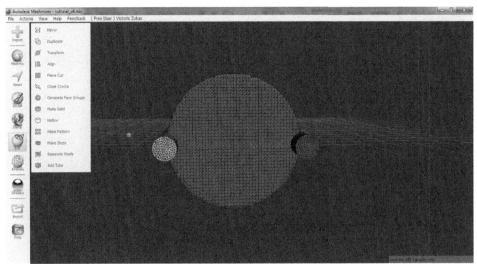

*Figure 5.44 -- Placement of both cylinders*

Duplicate the cylinder again, and position it on the top left of the handle.

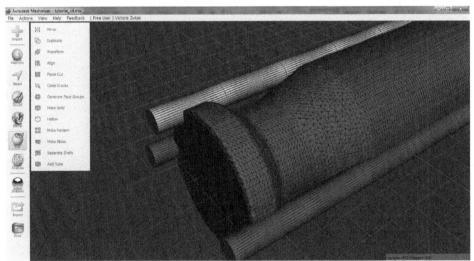

*Figure 5.45 -- Placement of the newest cylinder on the top left of the handle*

Duplicate the cylinder another time, and repeat the above step but placing it on the top right of the handle.

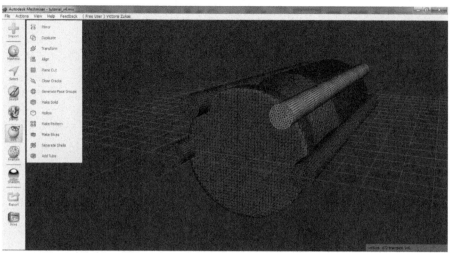

*Figure 5.46 -- Placement of the cylinder on the top right of the handle*

Repeat the above two steps, but move the cylinders so that they are placed on the bottom left and bottom right respectively.

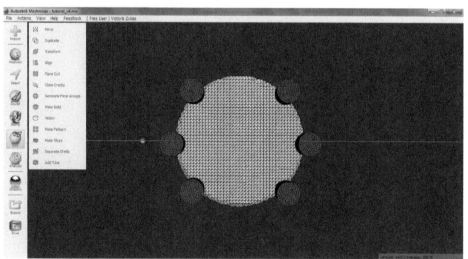

*Figure 5.47 -- Placement of the cylinders on the bottom right and bottom left of the handle*

Select one of the cylinders then, pressing Shift while clicking, select the other five as well. The Edit menu will change and new options will appear. On the menu select Combine. Now all of the cylinders are one object. Select the handle, and then click on one of the cylinders. All of the objects should be highlighted. It is VERY important that you select the handle first; otherwise this next step won't work right.

On the Edit menu, click on Boolean Difference and then Accept. Meshmixer will think for a moment, and then the cylinders will disappear and they will be replaced with new indents in the handle where they intersected with it. Now our handle is complete.

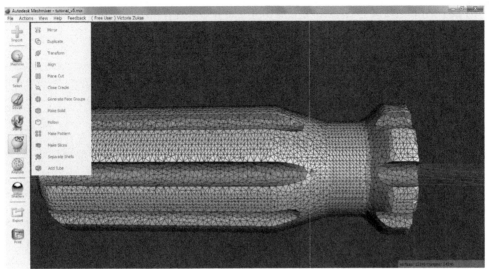

*Figure 5.48 -- The handle after using Boolean Difference*

If your handle didn't turn out like the picture above, then you need to reposition the cylinders. On the top menu bar click Actions, then Undo, twice (or press Ctrl + Z twice) to go back to before you selected Boolean Difference. Select the cylinders and then on the Edit menu click Separate Shells. This will break them back into separate objects. Move them either closer or further away from the handle (depending on how your handle turned out). Then recombine them and select Boolean Difference again.

There are two potential issues you could see if you didn't get it right on the first try: (Don't feel bad if you didn't. It took me a few tries to get it to look exactly the way I wanted.)

If you can see the inside of the handle after clicking Accept, you have holes in your mesh. The inside of the handle will look pinkish orange, so pan around and make sure you don't see any holes. If this happened, use the steps I mentioned above to fix it.

You might see parts of the mesh surrounded in a blue outline. This happens if the cylinder(s) weren't positioned correctly and some of the mesh of cylinder got left behind. You can use the above method to fix it but there is also a simpler solution.

Click on the Analysis icon and then Inspector. Anywhere that had that blue outlined geometry will have a purple line coming from it to a floating purple sphere. These spheres point out problem areas on the mesh. Click on the sphere, and Meshmixer will get rid of the offending geometry.

The picture below is what it would look like if you run into problem #2 and choose to use the Inspector method to fix it.

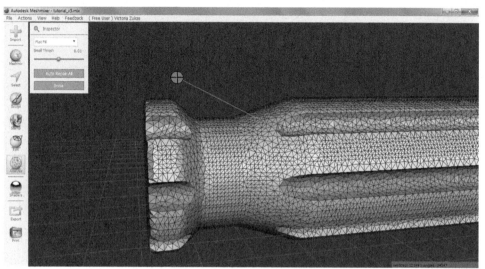

*Figure 5.49 -- Using the Inspector tool on leftover geometry*

Our handle looks perfect and we can move onto making the shank and the blade. Create another long thin cylinder, rotate it 90 degrees, and position it so that it sits in the middle of the handle. Make sure there are no gaps where the cylinder and the handle meet.

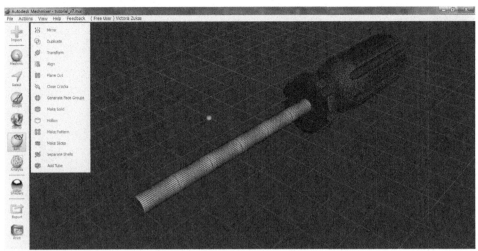

*Figure 5.50 -- The handle and shank of the screwdriver*

Click on Edit, and then Make Solid. Select Accurate from the drop down list. Set the Mesh Density to 60, the Min Thickness to 3mm and the Cull Edges Threshold to 20. After you make those changes click Update, and then Accept.

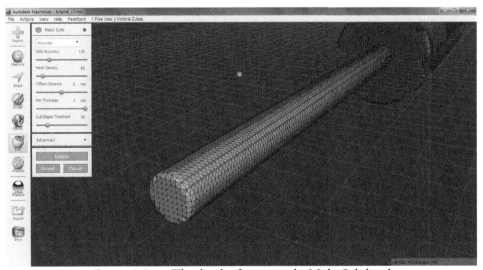

*Figure 5.51 -- The shank after using the Make Solid tool*

That's the shank; now for the blade. We're going to use Boolean Difference again to cut out the blade. Drag in a box from the Meshmix section. Use the manipulator tool to shrink it down so that the bottom is a little over two squares wide and four squares long on the grid. We will need two of these, so under the Edit menu select Duplicate and drag the new box off the side for now.

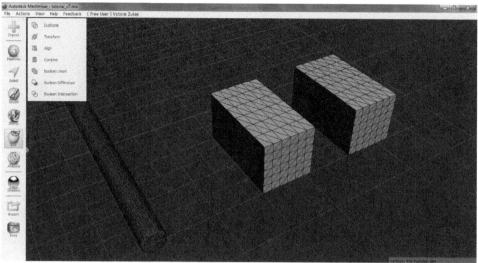

*Figure 5.52 -- Two new boxes added to the 3D View*

Move the first box so that it sits over the end of the shank. Title the box 10 degrees and move it up so that it looks like the picture below.

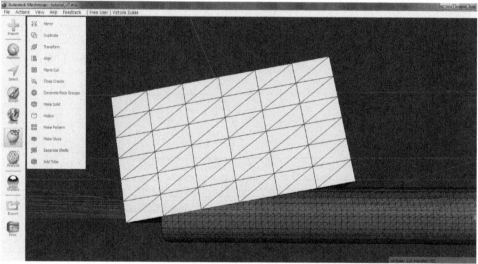

*Figure 5.53 -- Correct positioning of the first box*

Select the shank, then the box. Next click Boolean Difference, then Accept. You will now have half of the blade cut out of the shank.

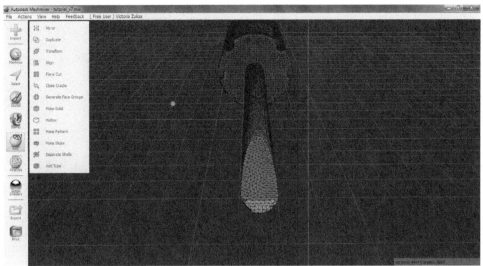

*Figure 5.54 -- Top half of the blade created by using Boolean Difference*

Position the other box you made on the bottom half of the shank. Rotate the box - 15 degrees and position it so that the right edge of the box is parallel with the edge of the top half of the blade. It should look like the picture below.

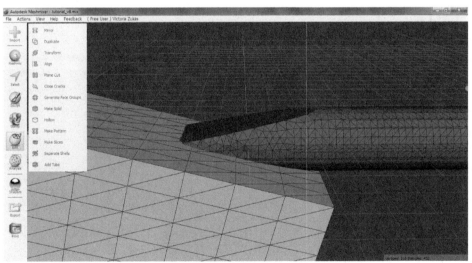

*Figure 5.55 -- Correct placement of the second box*

Click on the shank and then the box. Select Boolean Difference and then Accept. Now the blade is complete.

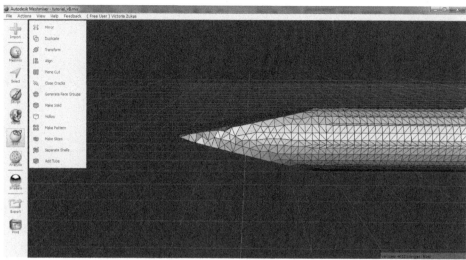

*Figure 5.56 -- The full blade created by using Boolean Difference*

The last step is to combine our objects together. Click on the shank and the handle. Under the Edit menu select Combine. The picture below shows the completed screwdriver. You can also clean up the 3D view by going into the Object Browser and deleting the original version of the handle and shank. **This is not a necessary step, and if you chose to do this make sure you save a new version of the screwdriver in case you need the deleted objects later.**

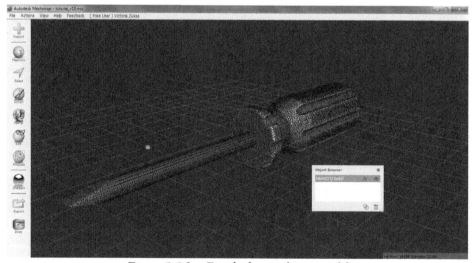

*Figure 5.56 -- Finished screwdriver model*

Congratulations! You're screwdriver is now complete. Now all that's left to do is print your new screwdriver. In the Print window select i.materalise as the print service. If you deleted the original handle and shank earlier then you don't need to follow the next step. If you didn't delete them, select the handle from drop down list of objects and click Remove from Cart. Do the same thing for the shank.

In the list of materials select any material from the Polyamide section. Polyamide can be used in multiple applications and is a good material to start with when 3D printing. Using the Polyamide material the model will cost between $17 - $26 UD, depending on which color and finishing options you choose.

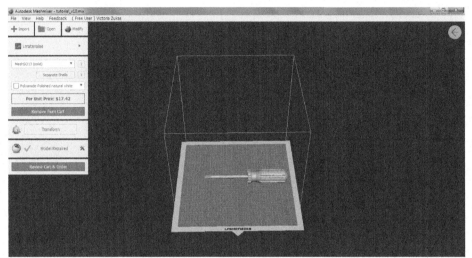

*Figure 5.57 -- The screwdriver model in the Print view. "Polyamide – Polished natural white" is the material selected.*

After purchasing your screwdriver from i.materialise you will receive a confirmation email. Within 24 hours you should be contacted by someone from customer service informing you that your screwdriver has been put in the printing que, along with the expected date the model will ship. Though it will be hard, now you just have to sit back and wait.

All that waiting will be worth it though. In the picture below you will see the final result of your efforts.

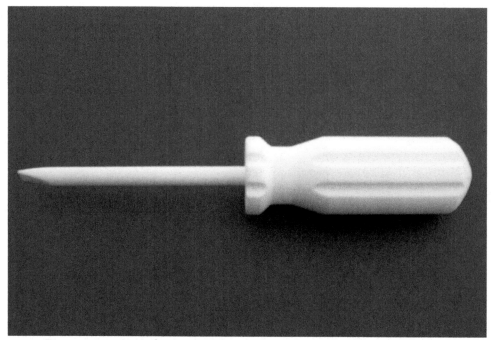

*Figure 5.58 – Screwdriver printed in "Polyamide – Polished natural white"*

## Further Reading and Viewing on Meshmixer

1. http://www.meshmixer.com/
2. https://www.youtube.com/playlist?list=PLu8TYSQ5jCFjdQBHsLoybhdKX OTmpTRlb
3. https://www.youtube.com/watch?v=aFTyTV3wwsE&index=17&list=PL5A 8E6C86E62A6F0C
4. http://i.materialise.com/blog/entry/3d-printing-with-meshmixer-a-beginner-friendly-introduction-to-3d-sculpting-and-combining-meshes

# Conclusion

Congratulations.

If you are new to 3D printing, and if we have done our jobs correctly, you now know about a 100 times more than you ever wanted to know. If you're in the field already, and we've done our jobs correctly, this book has served as a guide to the available techniques and associated literature for developing models for 3D printers.

All that's left is the final question: What's next?

Now is the time for you to use this information to further your specific interests in 3D printing. 3D printing is no longer a curiosity for those in engineering or computer science. It's the wave of the future, paving the way for advancements in every major industry. Improvements are being made to the technology every day. The possibilities are endless. At the moment we can only speculate how far 3D printing will go. Now it's your turn to be a part of this adventure. We wish you the best as you bring your creations and new ideas to life with 3D printing.

CPSIA information can be obtained
at www.ICGtesting.com
Printed in the USA
LVOW06s1447280617

539679LV00020B/172/P

9 781622 878963